13 Crystals

Growth, Properties, and Applications

Managing Editors: H. C. Freyhardt and G. Müller

Organic Crystals I: Characterization

Guest Editor: N. Karl

With Contributions by
H. Klapper, M. Kobayashi, T. Kobayashi, K. Sato

With 133 Figures

Springer-Verlag
Berlin Heidelberg New York
London Paris Tokyo
Hong Kong Barcelona Budapest

Managing Editors

Prof. Dr. H. C. Freyhardt, Kristall-Labor der Physikalischen Institute,
Lotzestr. 16–18, D-3400 Göttingen
and Institut für Metallphysik der Universität Göttingen
Hospitalstr. 12, D-3400 Göttingen, FRG

Prof. G. Müller, Institut für Werkstoffwissenschaften VI,
Universität Erlangen-Nürnberg, Lehrstuhl für Werkstoffe der Elektrotechnik,
Martensstr. 7, D-8520 Erlangen, FRG

Guest Editor

Prof. Dr. N. Karl, 3, Physikalisches Institut der Universität Stuttgart,
Pfaffenwaldring 57, D-7000 Stuttgart 80, FRG

ISBN-13: 978-3-642-76255-0 e-ISBN-13: 978-3-642-76253-6
DOI: 10.1007/978-3-642-76253-6

Softcover reprint of the hardcover 1st edition 1991

Typesetting: Thomson Press (India) Ltd., New Delhi;
51/3020-543210 – Printed on acid-free paper

Advisory Board

Table of Contents

Epitaxial Growth of Organic Thin Films and Characterization of their Defect Structures by High-Resolution Electron Microscopy

Takashi Kobayashi

Institute for Chemical Research, Kyoto University, Uji, Kyoto-Fu 611, Japan

1 Introduction

Lattice defects of organic molecular crystals affect their optical or electrical properties by changing the local energy structure. Lattice defects also play a very important role in the chemical and physical properties, for example, as an active site of a catalyst or an initiating point of a solid state reaction. However, very little has been reported on the defect structure of real organic crystals. In the past ten years it became clear that the origin and the structure of the defects depend on the geometrical and chemical nature of the building units of the crystal, the molecules. Molecular size, form and anisotropy, charge distribution, etc. cause the characteristic structure of the defect. Accordingly, a defect structure found in one compound may not be found in others. The defect structure of an organic crystal cannot be defined solely by the displacement of the molecular center from the normal lattice site. A rotational displacement of a molecule is frequently accompanied by a parallel shift of the molecular center. In addition to the usual geometrical crystallographic defects, *chemical* defects are important too which originate, for example, from differences in the substitution sites of molecules carrying side groups. In order to reveal such defect structures, direct imaging of molecules by high resolution electron microscopy is the only direct method. The purpose of this contribution is to introduce the real structures of the defects in some organic crystals and to try to interpret their origin from the point of view of anisotropic molecular shapes and adhesion forces. The crystals used here as examples are all composed of planar molecules which form a so-called columnar or "stacked" structure.

2 High-Resolution Electron Microscopy and Electron Diffraction

A modern electron microscope can resolve an interatomic distance as short as 1.2 Å and many data concerning local structural changes in inorganic materials have been accumulated by high-resolution electron microscopy. However, the first example of a *high-resolution* electron micrograph is the lattice image of an organic crystal, Pt-phthalocyanine (Pt–Pc) taken by Menter et al. [1]. In this chapter, the principle of the image formation in high-resolution electron microscopy is briefly surveyed and its application to several organic crystals described.

2.1 Principles of Electron Microscopy

Among a number of physical methods used for structure analysis of solid materials, only high-resolution electron microscopy combined with the electron diffraction method has the characteristics needed for a study of very small specimens of nanometer sizes and of nonperiodic structure such as molecular orientations at crystal defects, or even of an isolated single molecule. These characteristics are also available for the studies of crystal nucleation or of solid-solid or solid-vapor reaction processes. High-resolution electron microscopy is based on the same principle as X-ray structure analysis. The former method performs the Fourier-synthesis of the diffracted beams by the aid of electromagnetic lenses, while the latter does the synthesis by the use of a high-speed computer, as is illustrated in Fig. 1. The illustration shows the ideal case of the image formation process by a lens. For a real specimen in three dimensions, the projection along the beam direction is imaged as a two-dimensional picture. The electron diffraction pattern is formed in the back focal plane of the objective lens and is enlarged by one or two intermediate lenses and one projector lens for observation. In order to produce a diffraction spot, the lattice spacing (d) and the direction of the reflection, defined by the Bragg angle (θ), must satisfy the Bragg condition, $2d\sin\theta = n\lambda$, where λ is the wavelength. This relation can also be explained schematically by the Ewald construction, a geometric correlation between the reciprocal lattice of the crystal intersected by a sphere of radius $1/\lambda$ [2]. Figure 2 shows this relationship. The center of the Ewald sphere (C) is at a distance $1/\lambda$ from the origin (O) of the reciprocal lattice and the direction of the incident beam vector k_0 is $\overrightarrow{CO}$. The direction of the diffracted beam is given by the vector $k = \overrightarrow{CP}$ if a reciprocal lattice point (P) lies on the sphere. The vector $g = \overrightarrow{OP}$ is a reciprocal lattice vector and equal to 1/d. In the case of high-voltage electron diffraction the radius of the Ewald sphere $1/\lambda$ is usually very large as compared with typical reciprocal lattice distances because the de Broglie wavelength of the electrons is very short (for example, 0.025 Å for 200 kV and 0.0087 Å for 1000 k V). Therefore, the curvature of the Ewald sphere is so weak that a small area of its surface can be considered as a plane. Especially when the specimen is thin enough, (high resolution demands a specimen thinner than ca. 10 nm) the reciprocal lattice points elongate along the normal of the specimen

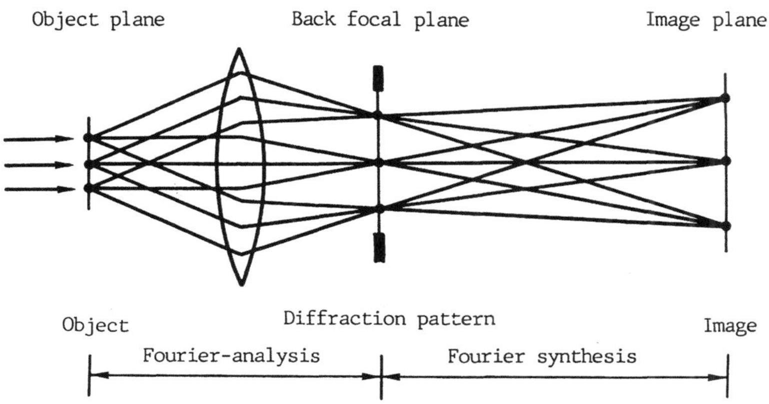

Fig. 1. Imaging processing by a lens or by a computer

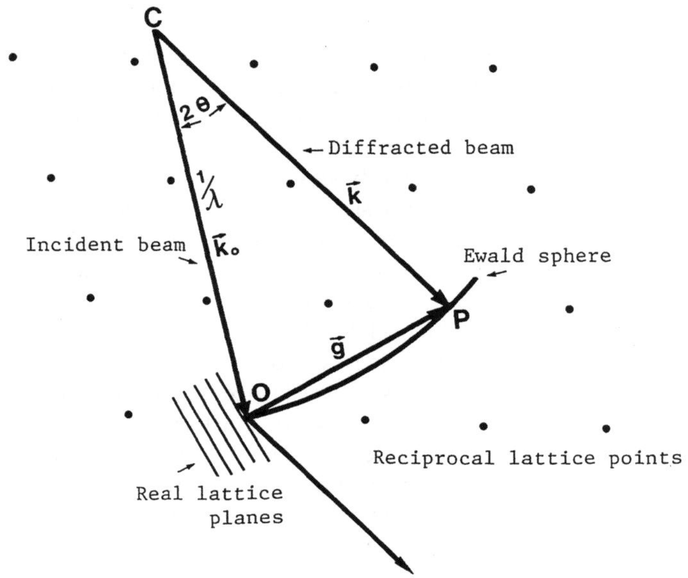

Fig. 2. Ewald sphere and reciprocal lattice points representing the Bragg reflection condition

thin film plane, with a length proportional to the inverse thickness of the specimen. Therefore, many reciprocal lattice points satisfy the Bragg condition at the same time and many diffraction spots can be recorded with the same sample orientation. The diffraction pattern recorded on photographic film, in a first approximation, can be interpreted in terms of an intersection of the reciprocal lattice with a plane including the origin of the reciprocal lattice, and thus gives information on the orientation of the specimen and other crystallographic properties.

The image contrast of an electron micrograph is produced by several mechanisms [3]. The diffraction contrast originates from the cutting off of some or all diffracted beams by an objective aperture in the back focal plane, or it is due to anomalous diffraction caused by a crystal defect. Figure 3a shows an example where some couples of equi-inclination fringes and the existence of defects can be observed. The mechanism causing the equi-inclination fringes is explained in Fig. 3b. The diffracted beams (hkl) and (hkl) are stopped by the aperture and thus do not arrive at the image plane. That part of the image which corresponds to regions in the sample where the planes contribute to the diffractions becomes unilluminated.

Within the crystalline surrounding near the defect, on the other hand, the diffracted beam deviates from the Bragg condition due to small lattice distortions and produces the contrast. By this method, dislocations and other defect structures of simple metals have been investigated in great detail in the past fourty years. However, this method is in most cases unsuitable for studying the

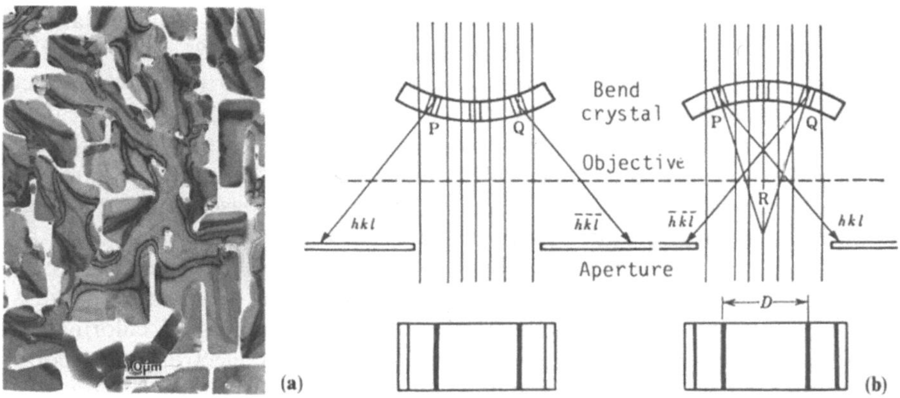

Fig. 3 a, b. Electron micrograph showing equi-inclination fringes (bent contours) of a perylene crystal (**a**) produced by cutting-off the diffracted beams by an objective aperture (**b**)

detailed defect structure of organic crystals, because of the complex shapes of the constituting molecules. Direct atom imaging by high-resolution electron microscopy, in contrast to X-ray diffraction techniques, is a powerful method for such investigations.

The contrast in a high-resolution image is produced by the phase shift of the scattered electrons due to the potential of the specimen [4]. If the specimen has the potential distribution $\phi(x, y, z)$, the image intensity $I(x, y)$ can be represented by:

$$I(x, y) = 1 + 2\sigma\phi(x, y) * s(x, y)$$

where σ is the interaction constant and equals to $\pi/\lambda E$, and $\phi(x, y)$ is a projected potential of the specimen along the z-axis $\phi(x, y) = \int_0^t \phi(x, y, z)dz$; where t is the thickness. The $*$ sign denotes a convolution operation and $s(x,y)$ is the spread function. This means that image contrast is given directly as the deviation of the intensity from unity which represents a projected potential of the specimen smeared out by the spread function $s(x, y)$. The spread function $s(x,y)$, which determines the resolution, is the Fourier transform of other aperture function $A(\theta)$ multiplied by the contrast transfer function $\sin \chi(\theta)$, where $\chi(\theta)$ is given by:

$$\chi(\theta) = \frac{\pi}{2\lambda}(C_s\theta^4 - 2\Delta f\theta^2)$$

This is equivalent to a multiplication of the amplitude of the diffracted beams on the back focal plane of the objective lens by $A(\theta) X \sin \chi(\theta)$. The $\sin \chi(\theta)$ is a function of the wavelength λ, defocus value Δf, spherical aberration constant C_s and the scattering angles θ, and determines the resolution. $A(\theta)$ is unity for the diffracted beam within the aperture, and zero outside the aperture. For a correct interpretation of the image contrast, the function $\sin \chi(\theta)$, which is usually called "phase transfer function", should be made close to unity over a region of the diffraction pattern as large as possible. Under such conditions many diffracted beams can recombine to form an image with the correct phases and amplitudes. The values of C_s and λ are intrinsic properties of each electron microscope, but the defocussing value Δf can be selected by the operator. The effects of defocussing on the image can be verified by studying the images of a sample whose structure is known. Figure 4 displays high-resolution molecular images of chlorinated-copper-phthalocyanine [5] taken with the Kyoto-500 kV electron microscope. These images can be obtained at the optimum defocus value, the so-called Scherzer focus, of $\Delta f = 450$ Å (under-focus). Copper atoms at each molecular center and chlorine atoms at the peripheries of the molecules are clearly discriminated as black dots. When the defocus values deviate from the optimum, the image features change remarkably as shown in Fig. 5a,b which are the images of the same specimen taken at an over-focus of $\Delta f = +200$ Å (a) and at an under-focus of $\Delta f = -1000$ Å (b). The corresponding transfer functions are shown in Fig. 6. The Scherzer focus is the optimum focus at which the highest resolution of the image, defined by $\delta_0 = 0.65 C_s^{1/4} \lambda^{3/4}$, can be obtained for each electron

Fig. 4. Molecular images of chlorinated copper phthalocyanine at the optimum focus (Scherzer focus)

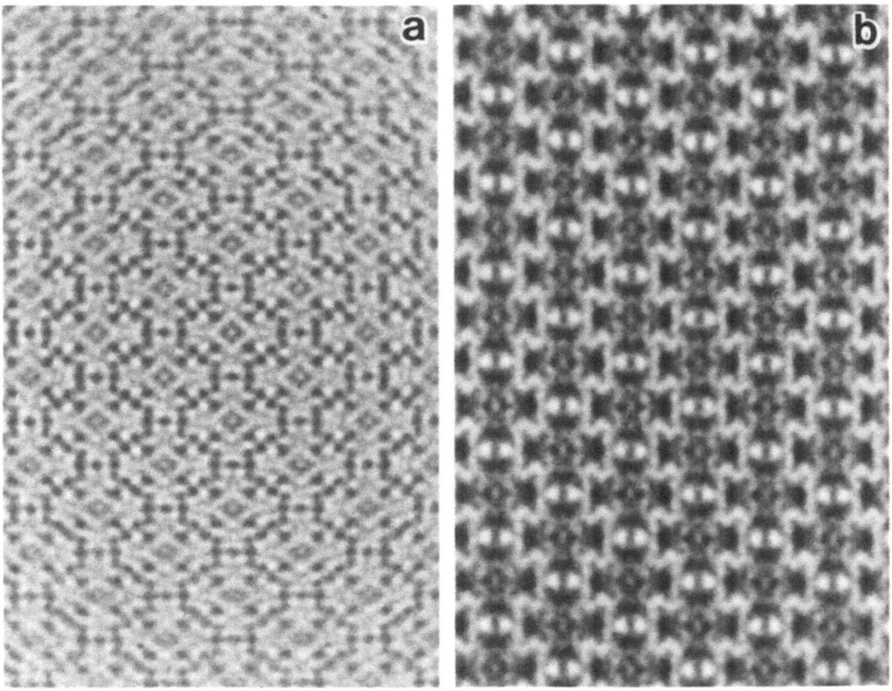

Fig. 5a, b. Changes of image contrast depending on the defocussing values. **a** over-focus ($\Delta f = +200$ Å), **b** under-focus ($\Delta f = -1000$ Å)

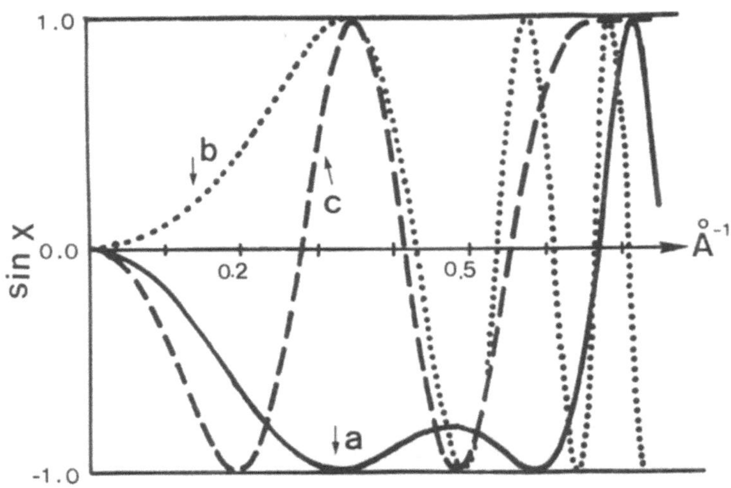

Fig. 6a, b. Transfer functions corresponding to the individual images shown in Figs. 4 and 5. (**a** optimum-focus, **b** over-focus, **c** under-focus)

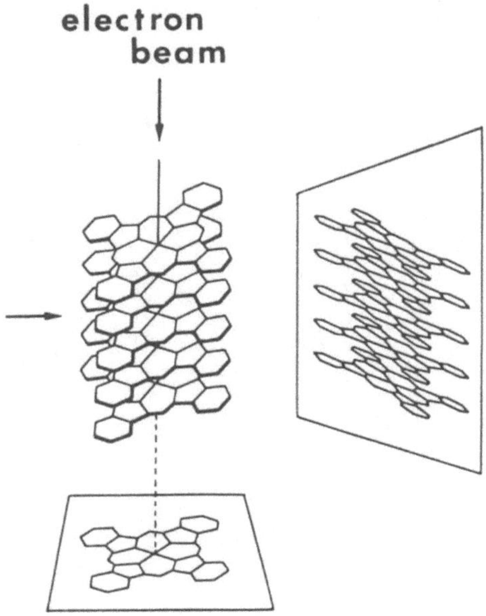

Fig. 7. A molecular column of planar molecules and its detectable projected images. The side view of the molecules is imaged as a streak only due to a lack in resolution

microscope. Evidently the images change as a function of the defocus values and an intuitive interpretation of an image is possible only at the Scherzer focus. This approach is valid for thin specimens which can be considered as weak-phase objects. Otherwise, the image contrast and potential distribution in the specimen are combined in a nonlinear way. The molecular images shown here are not an image of a monomolecular film but of a single crystal whose thickness is about 50 Å. Therefore, ten or more molecules are stacked in columns and the images correspond to the projections of the molecular columns along the column axis, as illustrated in Fig. 7. Thus a parallel alignment of the crystal zone axis to the electron beam direction is also important in order to obtain a simple structure image, and for that purpose a goniometer stage is inevitably needed in the electron microscope, in addition to a preparation of the specimen in a selected suitable orientation (see below).

2.2 The Resolution Limit

The theoretical point-to-point resolution (δ_0) is limited by the resolving power of the objective lens according to the general rule of optics known as the Abbe [6] formula, $\delta_0 = 0.65\lambda/\sin\theta \simeq 0.65\lambda/\theta$, where θ is the scattering angle of the diffracted electrons, which is usually in the region of 30 mrad. The formula indicates that if one wants to obtain finer information, one must increase the acceptance angle in the objective aperture for the scattered waves as much as possible. However, owing to the existence of lens aberration, especially spherical aberration of the objective lens, the waves with higher scattering angles suffer from increasingly more distortion in their phases, depending on the scattering angle θ and defocus value of the lens as described in the previous section. It is worthwhile to note that a higher accelerating voltage, which gives a shorter wavelength, increases the resolving power, but only if C_s is constant and if high voltage can be stabilized at the same level as the lower voltage. The theoretical resolution limit is thus deteriorated by unstabilities of the high voltage as well as by incorrect illumination conditions.

However, for the case of organic specimens the practical limit of resolution is set by radiation damage which the specimen suffers owing to the intense electron irradiation needed for obtaining a highly magnified image. The extent of the damage is first detected as a deterioration of the crystalline state and a degradation of the image quality. According to the Rose equation, Glaeser et al. have derived a formula from which one can obtain the attainable resolution (δ) for each material [7]:

$$\delta \geqq \frac{k}{c\sqrt{fq}}$$

where q is the critical electron dose, defined as the electron dose which diminishes

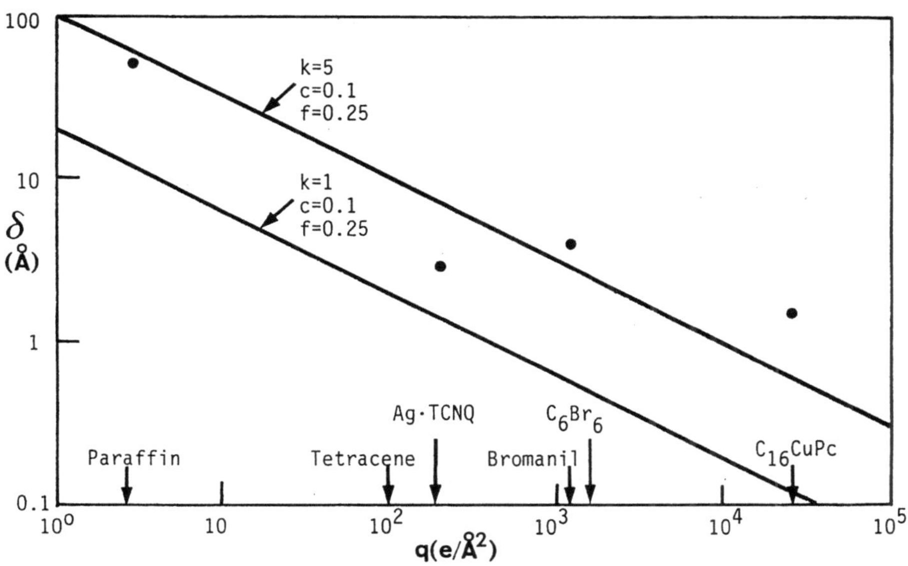

Fig. 8. The radiation damage-limited resolution (δ) of different organic specimens (observed examples)

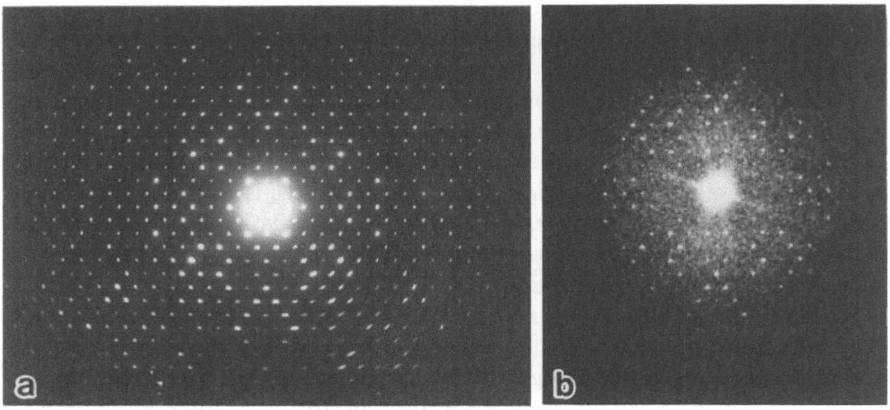

Fig. 9 a. Electron diffraction pattern of the specimen shown in Figs. 4 and 5. **b** Optical laser diffraction pattern of the electron micrograph shown in Fig. 4. The diffuse background in b represents the transfer function shown in Fig. 6a

the electron diffraction intensity to 1/e of its initial value, c is object contrast and usually taken as 0.1, f is the utilization factor or detection efficiency (0.25).

The ultimate resolution attainable has been reported for some organic materials; examples are shown in Fig. 8. In order to use the permissible dose for each specimen as efficiently as possible, the images should be recorded by applying the minimum dose system developed by several authors [8]. One such device was made in our laboratory and proved efficient in recording molecular images. The device allows focussing on a small spot area close to the target area to be finally imaged; the latter is subsequently exposed to the total permissible illumination for image recording. Whether the recorded image retains the finest spatial information as expected, or not, can be checked by observing the photodiffraction pattern generated by laser illumination of the photographically recorded image [9]. Whenever the specimen is thin enough to be treated as a thin phase object, the laser diffraction spot pattern thus obtained corresponds to the (reciprocal) lattice spacing retained in the photograph of the electron image, and the resolution limit, set by degradation due to the radiation damage can be judged. The optical diffraction pattern is also useful for the determination of the local crystal orientation, as well as for estimating the defocus value during image recording. Figure 9 displays the electron diffraction pattern belonging to Fig. 4(9a) and the laser diffraction pattern of the image shown in Fig. 4(9b).

2.3 Determination of the Sample Orientation by Electron Diffraction

As mentioned in Sect. 2.1, the electron diffraction pattern of a very thin crystalline film can be interpreted as a section of the reciprocal lattice by the Ewald sphere,

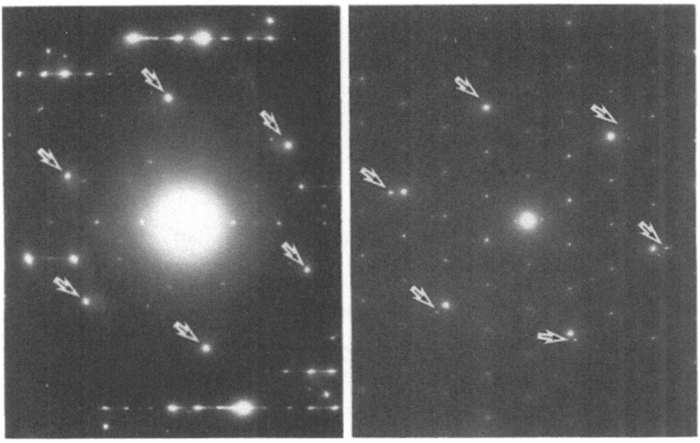

(a) (b)

Fig. 10 a, b. Electron diffraction from a C_6Br_6 epitaxial film in *n*-orientation on MoS_2. The crystal zone axis of C_6Br_6 is tilted with respect to the electron beam in (**a**) and is parallel in (**b**). The hexagonal net pattern (arrowed spots) from the substrate MoS_2 is distorted in **b**

and in case of high velocity electrons the radius of the sphere is so large compared to the repeat unit of the reciprocal lattice that the section of the Ewald sphere, which gives rise to the observed diffraction pattern, may be assumed to be approximately planar. (The finite curvature of the sphere must be taken into consideration, however, whenever a *precise* value of an interplanar distance shall be determined from the diffraction pattern).

An epitaxial film of hexabromobenzene grown on the (001) plane of molybdenite, MoS_2 (cleaved in air) gives a diffraction pattern as shown in Fig. 10a. The diffraction spots marked by arrows which are arranged in a hexagonal pattern are produced by the substrate, while the other spots are from C_6Br_6. The latter spots are strong along one direction in the figure and do not form a simple net pattern. This means that there is no principal axis of the crystal C_6Br_6 aligned parallel to the beam direction. When the crystal was slightly rotated the pattern changed markedly and became simpler as shown in Fig. 10b. In this latter pattern, diffraction spots from the MoS_2 deviate from the original exact hexagonal pattern, while the spots from C_6Br_6 from a regular net pattern displaying the a*–c* reciprocal plane. By analysis of the deviation of the spots of MoS_2 from their original arrangement, the tilting angle and the direction of tilting can be determined. This situation is explained in Fig. 11. The molecules of C_6Br_6 form molecular columns by parallel stacking and the column axis coincides with the *b*-axis. The plane normal of the molecule makes the angle of 23° with the *b*-axis [10]. These facts can be interpreted by *epitaxial* growth of the organic layer on the inorganic substrate, as will be discussed later in more detail. The first molecules of C_6Br_6 are adsorbed flat on the MoS_2 and the columnar growth occurs from the first adsorbed molecules which act as nuclei. As a result the *b*-axis is tilted by an angle of 23° with

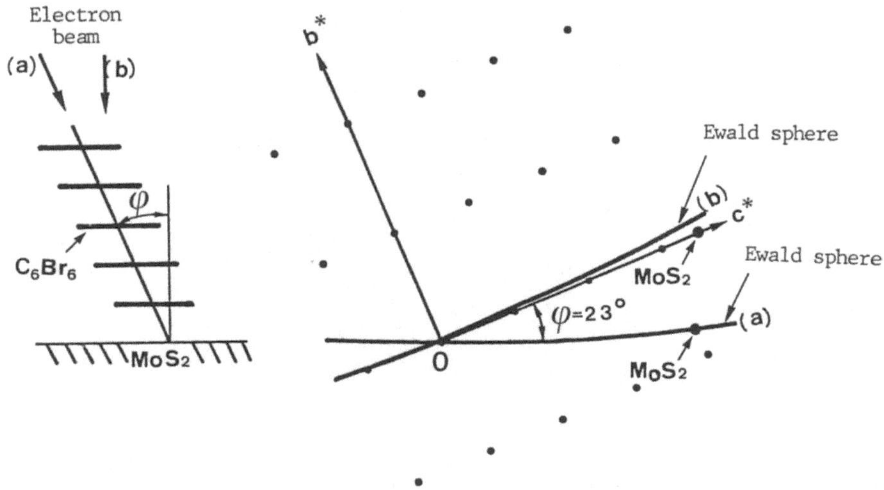

Fig. 11 a, b. The relation between the reciprocal lattice of C_6Br_6 and the Ewald sphere for two different angles of the incident electron beam. The reciprocal lattice points of MoS_2 are represented by arrowed dots; cf. Fig. 10

respect to the plane normal of the substrate. This adsorption is strong and dominates the normal cohesive forces which determine the usual mutual arrangement of the molecular columns in the regular crystal. Therefore, displacement of columns from their regular position in the ordinary crystal packing takes place along the column axis by fractions of an intrastack lattice constant. These defects, one-dimensional stacking faults, cause the diffuse scattering of the electrons and lead to the numerous streaks appearing in the diffraction pattern shown in Fig. 10a.

A crystal of TTF:TCNQ is composed of *separate* molecular columns of TTF and TCNQ, arranged in a "segregated stacking" pattern; the column axis is along the *b*-axis [11]. The crystal belongs to the monoclinic space group $P2_1/c$ and has the lattice constants a = 12.298 Å, b = 3.819 Å, c = 18.468 Å, $\beta = 104.46°$. Films of this charge-transfer (CT)-complex, vacuum-deposited onto a KCl (001) cleavage plane, are composed of microcrystallites oriented in two directions which cross at right angles (Fig. 12)[12]. In the high resolution electron diffraction tilted about the *b**-axis, the *a**-axis comes to fit the Ewald sphere and on the two main axes the spots form arrays with intervals of 1/12.29 $Å^{-1}$ and 1/3.78 $Å^{-1}$. Many crystals contribute to the HRED pattern, but the selected area electron diffraction (SAED) conveys the structure information from a small region of the specimen. The SAED showed that the reciprocal lattice units are 1/12.29 $Å^{-1}$ and 1/3.78 $Å^{-1}$, which reveals that the HRED in Fig. 13a is composed of two sets of similar patterns. When the crystal was tilted around the *b**-axis in the HRED, the features changed gradually. Near a tilting angle of 12°–15° the intensity distribution of the spot array on the equator, which is

Fig. 12. Electron micrograph of an epitaxial film of TTF·TCNQ grown on KCl. Film thickness is about 600 Å

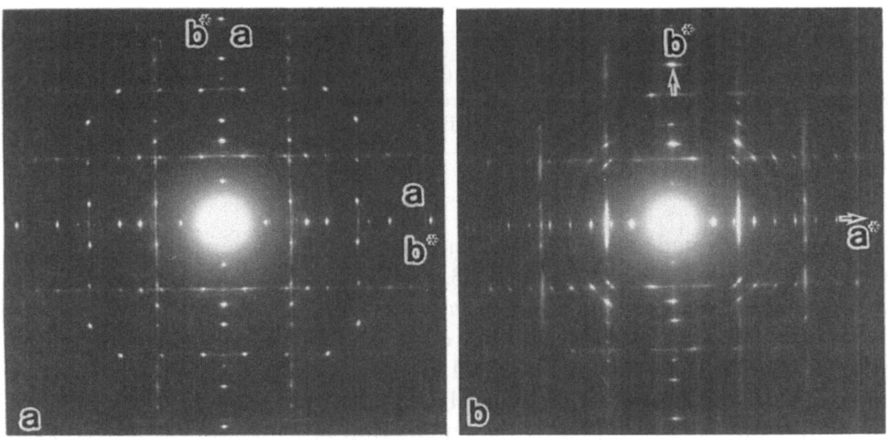

Fig. 13 a, b. HRED pattern from the thin film shown in Fig. 12 **a**, and pattern when the film was tilted 14°, about the vertical b^* axis **b**. The intensity distribution and the spacings on the equator change significantly by tilting

perpendicular to the tilting axis, indicated that another main axis in the reciprocal lattice came to tangenially touch the Ewald sphere and thus fulfill the diffraction conditions. The HRED with a tilting angle of 14° is shown in Fig 13b. The periodicity of the spot array is $1/11.9\,\text{Å}^{-1}$ and therefore the equator is assumed to represent the a^*-axis of the TTF:TCNQ crystal at this angle. This

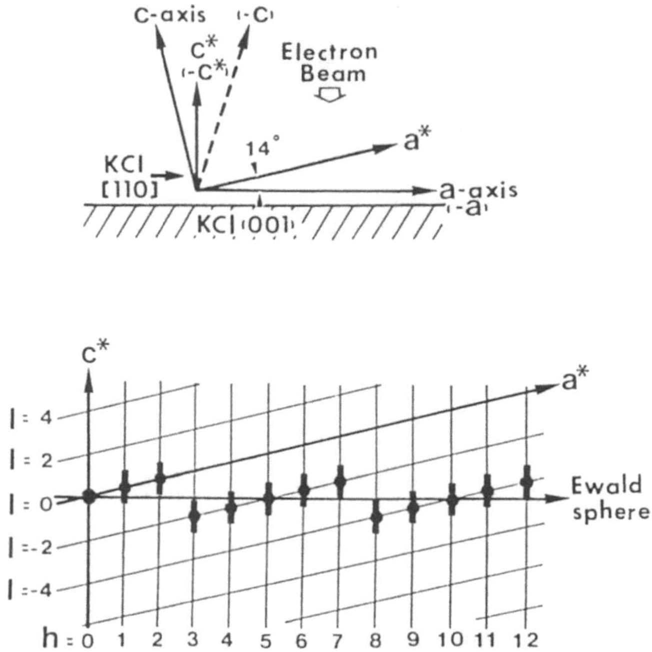

Fig. 14. The relationships between the real and reciprocal lattices of TTF·TCNQ on KCl. The direction of the *a*-axis coincides with one of the [110] axes of KCl. The Ewald sphere intersects at first the lower layers 1 = − 2 and 1 = − 4 of the reciprocal lattice; only after tilting it fits with a*. These situations correspond to Figs. 13a and b, respectively

situation is illustrated in Fig. 14. The reciprocal unit dimensions derived from the X-ray data are $1/11.909\,\text{Å}^{-1}$ and $1/3.810\,\text{Å}^{-1}$ for a^* and b^*, respectively. However, the lattice constant along the *a*-axis is 12.298 Å and this value coincides well with the inverse of the periodicity of the spot array on the equator of the original HRED pattern shown in Fig. 13a. Accordingly, it is supposed that the appearance of the inverse of the *a* unit cell length of TTF:TCNQ as the interval of the diffraction spots is caused by the intersection of the Ewald sphere with the *elongated* reciprocal lattice points, as is illustrated in Fig. 14. The elongation of the interference region around each reciprocal lattice point is due to the smallness of the dimension along the depth direction of the film. The a^*-axis is tilted by 14° with respect to the *a*-axis which is parallel to the substrate plane. When the film is tilted about the b^*-axis, the a^*-axis comes to fit the Ewald sphere and diffraction spots of [h00] appear, as actually observed in the HRED pattern of Fig. 13b. In conclusion, the TTF:TCNQ epitaxial film grows on KCl (001) with its *ab*-plane parallel to the substrate surface. The unit cell dimension along the *b*-axis is slightly shorter than that of the single crystal reported by Kistenmacher et al. [77].

2.4 Images of Organic Crystals at Molecular Resolution

The highest resolution of present-day electron microscopes reaches 1.2 Å in special cases; with a usual commercial microscope one can obtain 1.8–2.0 Å resolution. However, the carbon-carbon distance in molecules is shorter than 1.5 Å and cannot be resolved at present with a commercial one. In addition to the limited resolving power, the contrast given by light elements is low. Therefore, what can be derived unambiguously with organic specimens is only the arrangement of the constituting *molecules*. If there exist heavy atoms in the molecules the situation improves somewhat. Lattice images of the metal:TCNQ complexes, Ag:TCNQ [13] and K:F_4-TCNQ [14] are compared in Figs. 15a, b. In F_4-TCNQ the four hydrogen atoms of the quinone ring of TCNQ are substituted by fluorine atoms. The images show the effect of the substitution as an apparent increase of the width of the quinone rings of F_4-TCNQ, because of the higher contrast caused by the fluorine atoms as compared with hydrogen. Chlorine gives even higher contrast and is clearly discernible as shown in Fig. 4, while the unchlorinated molecule (Fig. 54) appears only barely resolved. Therefore, it is premature to discuss bond lengths on the basis of electron microscopic observations. All molecular images observed so far are not those of isolated single molecules but always projections of several molecules stacked along the electron beam direction. The molecular images therefore are a

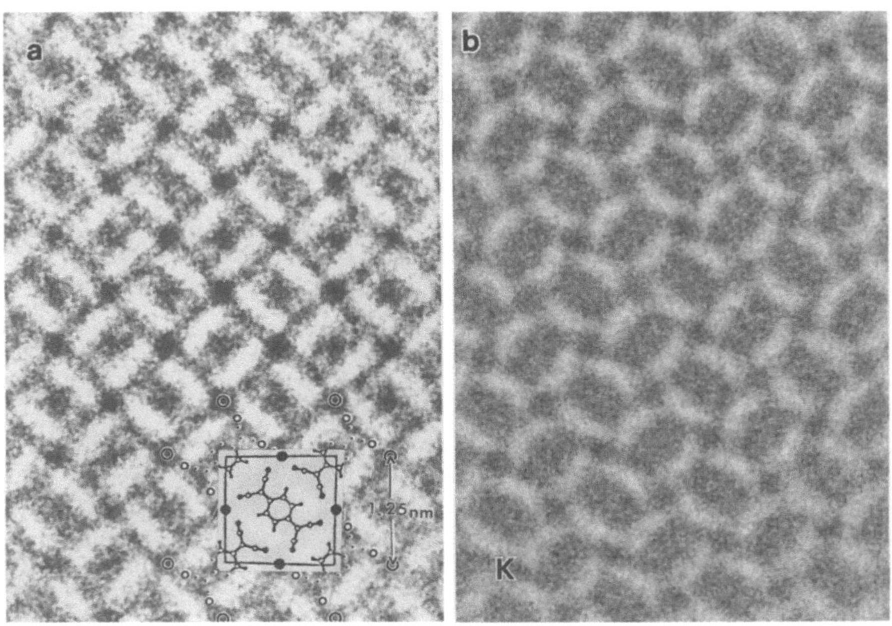

Fig. 15 a, b. Molecular images of Ag:TCNQ (**a**) and K:F_4-TCNQ (**b**) charge transfer complexes formed by a solid state topotactic reaction. The crystal structure of the former complex is shown in Fig. 36

projection of molecular columns; they can only be obtained if there are columnar stacks in the crystal structure and if the crystal is appropriately aligned to look along these in the electron microscope.

3 Preparation of Thin Films with Definite Orientations

As mentioned in the previous section, high-resolution electron microscopy requires very thin samples (30 ~ 100 Å). Vacuum deposition of the material is the best method to obtain a sufficiently thin specimen. When a material is vacuum-deposited onto an amorphous substrate, it generally forms a polycrystalline film in which each crystallite takes a random orientation. Therefore, an electron diffraction pattern of the film shows Debye–Scherrer rings. An example of such a case is displayed in an electron micrograph in Fig. 16a,b [15]. The sample is a Zn-phthalocyanine (Zn–Pc) film deposited in a high-vacuum onto a thin amorphous carbon film which was made by an indirect deposition method. At least three different kinds of images including polymorphs were found in the film. If one wants to make a film of an organic substance with a definite orientation of the crystallites, one should use an epitaxial growth method on a crystalline substrate or a topotactic formation of a reaction product from a single crystal or from an epitaxial film.

3.1 Apparatus

Several methods have been developed and used for the preparation of organic thin films. In this section the principles and techniques for obtaining a film of definite orientation by using the epitaxial or the topotaxial growth method are

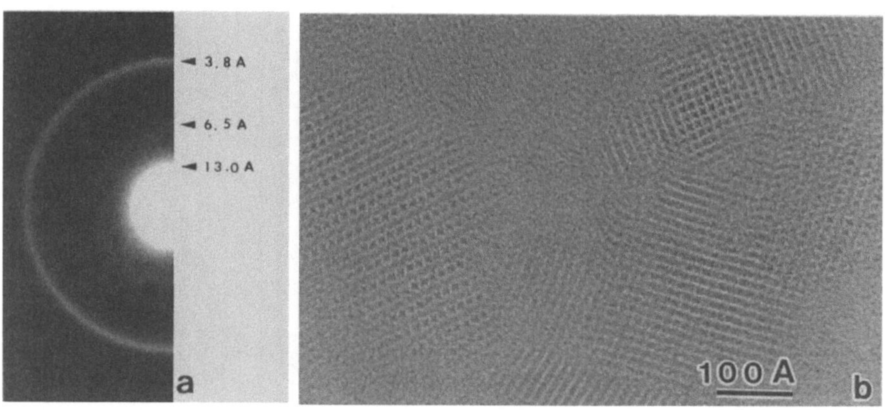

Fig. 16 a, b. An electron diffraction pattern (**a**) and an elecron micrograph (**b**) of a Zn–Pc thin film deposited on an amorphous carbon film

described in detail for some aromatic organic compounds. It is characteristic of topotactic reactions that the orientation of the crystalline reaction product is governed in most cases by the crystal structure of the original crystal and additionally influenced in several cases by specific interactions between the substrate and the molecules in the film. The epitaxial growth itself is a phenomenon well known since 1836 [16], when the occurrence of an oriented deposition from solution or vapor onto a different crystalline substrates was observed in crystal growth. Crystal formation from the vapor phase in general needs a high-vacuum chamber, equipped with a substrate holder which can be heated. A cold-finger is also useful for materials of low molecular weight which have high vapor pressure. An evaporation source, two sets of thermocouples for temperature measurements of the substrate and of the source, and a thickness monitor are required. A vacuum of 1×10^{-2} Pa is sufficient in most cases, but a vacuum higher than 1×10^{-4} or even 1×10^{-6} Pa turns out to be better for reproducible results. Ultra-high vacuum better than 1×10^{-7} Pa is recently used by some authors, but it is not yet clear how this influences the structure of the resulting films.

The substrate holder should be able to hold two or four thin slices of a substrate crystal material and to allow heating up to temperatures of about 600°C. The heater must be covered and shielded in order to avoid contamination and unnecessary heating of the chamber. A tungsten or molybdenum boat is commonly used as an evaporation source, but in our laboratory we use a small fused silica crucible inserted into a spiral wire heater as shown in Fig. 17. In cases where the temperature of the evaporating source has to be controlled, a crucible of the type used as a sample cell in differential thermal analysis, which has a small

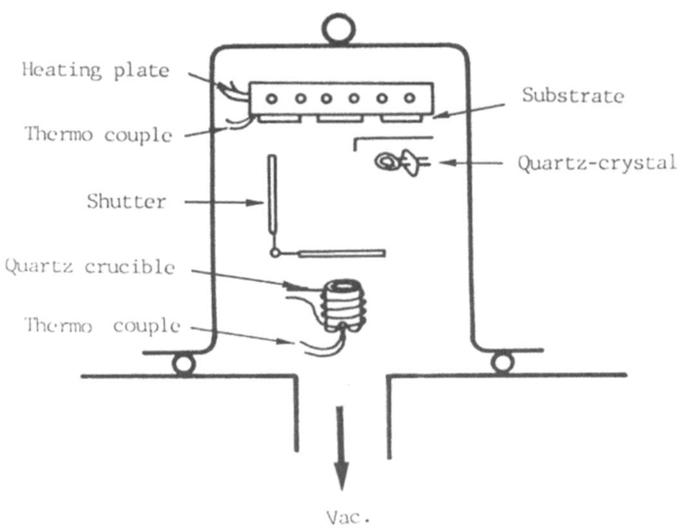

Fig. 17. Vacuum chamber used for the preparation of the epitaxial films exhibited in this article

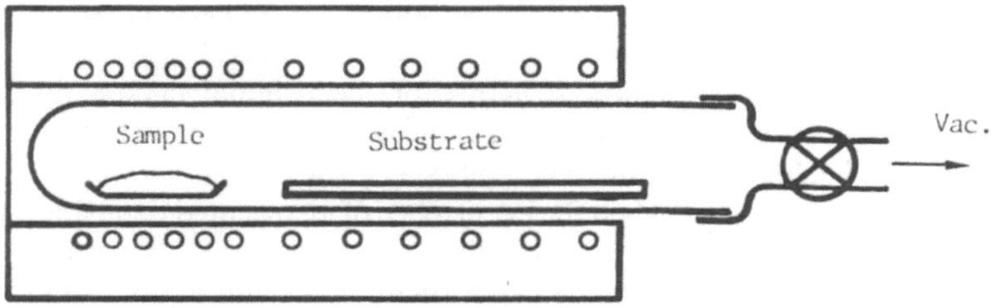

Fig. 18. Apparatus for using the sealed-tube method to obtain epitaxial films. The furnace should be temperature controlled

bore at the bottom for insertion of a thermocouple, is adequate. A Knudsen cell is also useful. The film thickness can be measured and controlled by a quartz oscillator micro-balance or by monitoring light-absorption. The former is convenient and becomes more and more popular, but appropriate heat-shielding of the quartz crystal is important. The measurement of the surface temperature of the substrate crystal is difficult, but very important. For that purpose it is convenient to use a thermocouple spot-welded to a small piece of platium sheet metal and to press it with a small stainless-steel nail to the substrate crystal. A shutter is very useful. At the beginning of evaporation of an organic substance, some more volatile impurities may evaporate in addition to thermal degassing; therefore, the pressure of the chamber rises. The shutter is opened only after stabilization of the pressure and the source temperature.

Another method of obtaining epitaxial films is to place the organic material (in a ceramic boat) and some pieces of substrate single crystals together in a tube which is subsequently evacuated and sealed as shown in Fig. 18. The tube is heated in a tubular furnace over which a temperature gradient has been established. If no such gradient is available, the temperature is raised a little higher than the sublimation point of the material and lowered very slowly down to room temperature. By this sealed tube method, the film thickness is controlled only by the amount of the material placed in the boat and by the size of the tube. The advantage is the possibility of obtaining a larger film area as with molecular beam deposition in the vacuum chamber.

3.2 Substrate

Many available single crystals can be used as substrates for epitaxial growth of organic materials. It should be noticed, however, that some organic materials react with the surface atoms or ions of the substrate [14]. Alkali halides are most popularly used for the epitaxial growth of metals as well as of organic substances. Rock salt (NaCl), sylvine (KCl), potassium bromide (KBr), and potassium iodide (KI) have the rock-salt type structure; they can easily be cleaved along the (100)

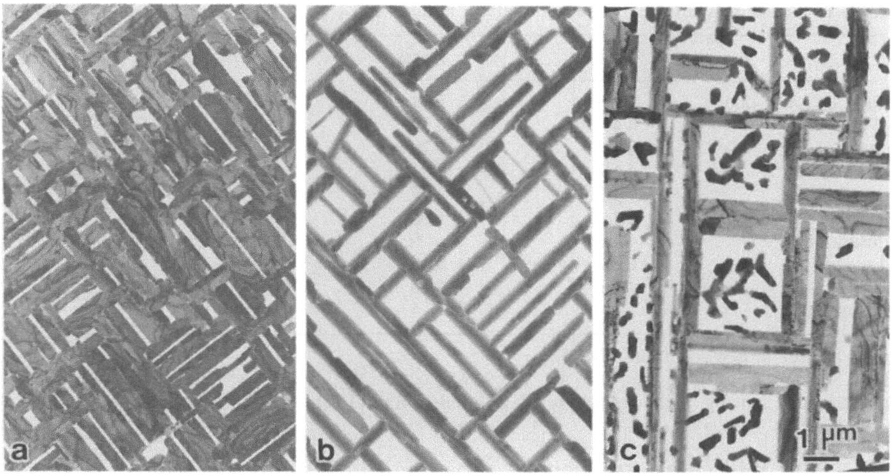

Fig. 19 a–c. Epitaxially grown hexabromobenzene showing the substrate influences of KCl (**a**) NaCl (**b**), KBr (**c**) on the thin film morphology. On KBr *n*-orientation appears together with *p*-orientation

plane. In addition to the lattice constants, the chemical nature characterizable by such parameters as ionization potential or electron affinity of each composing element, or binding energy of ions in a crystal, are important factors influencing the geometric structure of the film grown on their surface. The differences in crystal habits of hexabromobenzene grown on KCl, KBr, and NaCl cleaved faces can be seen in a series of electron micrographs shown in Fig. 19a,b,c. These substrate crystals were cleaved in air and mounted on the heating plate shown in Fig. 17. After two hours in a vacuum of 2×10^{-5} Pa without any heating of these crystals, pure C_6Br_6 was evaporated from a quartz crucible and condensed on them at room temperature. Crystallites of C_6Br_6 belong to a monoclinic space group and have the lattice constants a = 15.382 Å, b = 4.002 Å, c = 8.377 Å and $\beta = 92.7$ Å [10]. They are elongated along the two symmetry-equivalent directions, [110] and [$1\bar{1}0$], of each substrate. SAEDs from these crystallites reveal their habits and show that the preferred growth occurs under development of their (100) and (001) facets. From these morphological features follows that the C_6Br_6 molecules are placed with the molecular planes almost perpendicular to the substrate surface and therefore, the interaction between the molecule and substrate ions is not very strong. The differences in the crystal habits appear as different extensions of the habit plane in the lateral direction, as can be seen in the micrographs of Fig. 19. The crystals grown on KBr display remarkable differences from those grown on the other substrates. The lattice mismatches in each case are collected in Table 1 which shows that the mismatch is minimum for the case of NaCl and maximum for KBr, but the differences are not so remarkable for the other cases.

The molecular arrangement shall be considered in the next step. The full crystal structure of the C_6Br_6 is reported in Ref. [10]. For the orientation derived

Table 1. Lattice mismatches between the alkali halides and C_6Br_6

	KCl	NaCl	KBr
Lattice Const. (a_0)	6.29(Å)	5.64	6.59
d[110]	4.45(Å)	3.99	4.66
d:a	0.29	0.26	0.30
(d:a) × 4	1.16	1.04	1.20
mismatch %	16	4	20
d:b	1.11	1.00	1.16
mismatch %	11	0	16
d:c	0.53	0.48	0.56
(d:c) × 2	1.06	0.96	1.12
mismatch %	6	−4	12

a,b,c are the lattice constants of C_6Br_6,
a = 15.36 Å, b = 4.01 Å, c = 8.36 Å, $\beta = 92.65°$
d is a lattice spacing of the respective alkali halide along [110]

above, the molecular arrangement shown in Fig. 20 can be obtained. No large mismatch can be found in any case if it is assumed that the Br atoms of the C_6Br_6 molecules come close to the alkali ions.

When the substrate crystals were preheated at about 500°C in vacuum for two hours before the C_6Br_6 was deposited at room temperature, the crystallites composing the film remarkably altered their appearances with respect to the situation in the previous non-preheated case. The overall behavior is the same as in the previous case for KBr. But, after a heat pretreatment of the substrate the planar molecules tend to make more intimate contact to the surface in that they adsorb flat already in the initial stage of growth. Each adsorbed molecule acts as a nucleus. Thus the molecular arrangement, as well as the lattice mismatch are important factors for the epitaxial growth mode. This is confirmed, for example, by the micrographs shown in Fig. 21 where C_6Br_6 grown on preheated molybdenite (MoS_2) is shown. The molybdenite has a hexagonal symmetry which seems to fit well with the shape of the molecule of C_6Br_6. The lattice fit is also good in this case, as illustrated in Fig. 22. Therefore, the film of C_6Br_6 on MoS_2 is uniform, in contrast to the films grown on alkali halides shown above which were composed of microcrystallites (cf. Figs. 19,20).

When the interaction between the atoms or ions of the substrate and the guest molecules is too strong, chemical reaction can take place, e.g. under formation of a complex film. The K:TCNQ complex is a good example of such a case. When TCNQ is vacuum-deposited onto KI (001), it reacts and gradually produces a well oriented K:TCNQ crystalline film (Fig. 23), while on KCl it forms no complex but grows epitaxially. F_4-TCNQ [14], which is a stronger electron acceptor than normal TCNQ, reacts directly on KI under complex formation, and growth of the complex crystal takes place as soon as the deposition of molecules from the vapor phase occurs. This case can be named "epitactic reaction".

Muscovite (mica) is a popular and convenient substrate, because it is easily cleaved and relatively large slices can be obtained. Muscovite belongs to the

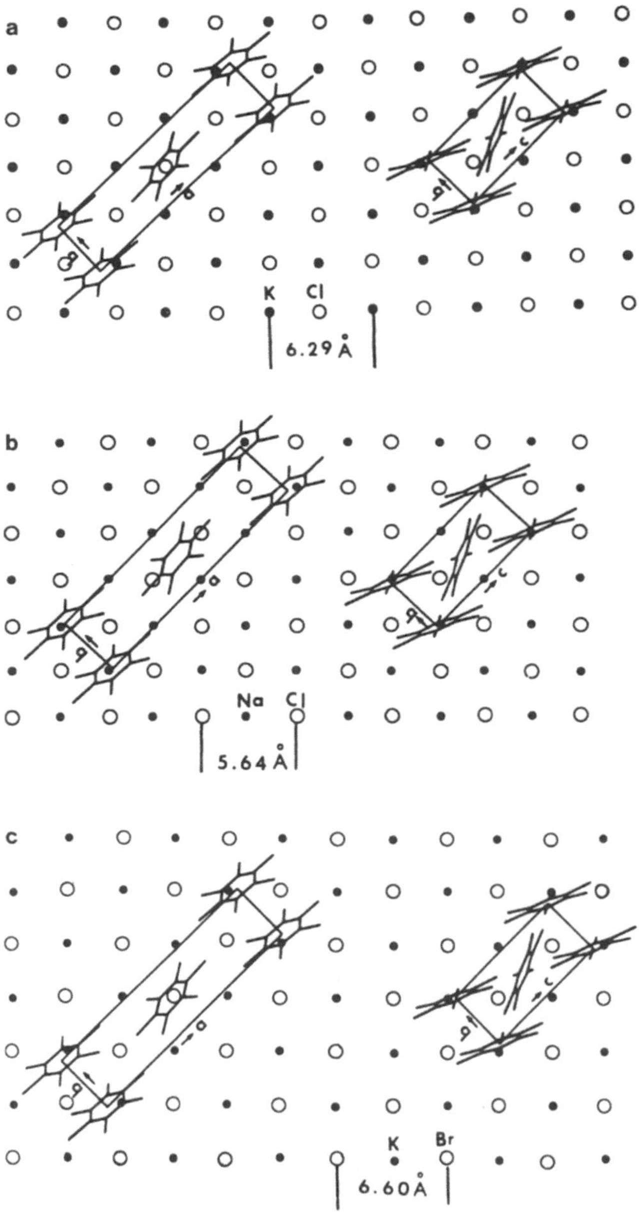

Fig. 20 a–c. The relation between the substrate (KCl, NaCl, KBr) and the epitaxially grown C_6Br_6 crystal lattice, showing the degree of lattice match. The epitaxy-relations are derived from diffraction data obtained from the samples shown in Fig. 19

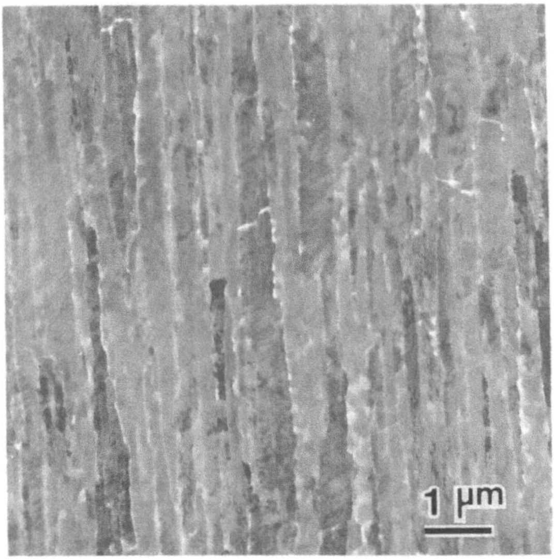

Fig. 21. Hexabromobenzene, C_6Br_6, grown on MoS_2

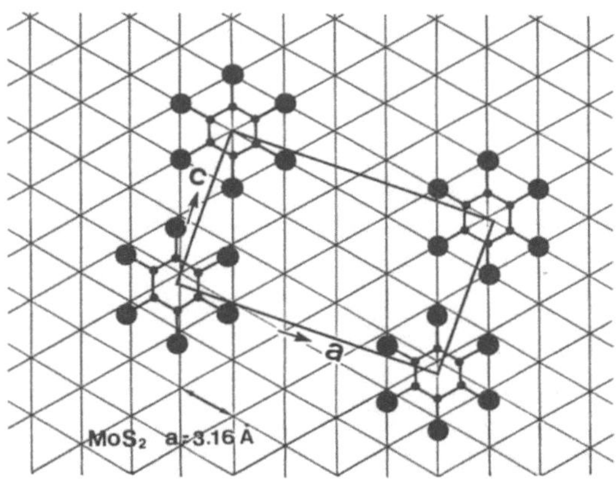

Fig. 22. C_6Br_6 on MoS_2. Lattice mismatch is small and molecular fitting to the substrate is good

monoclinic crystal system [17] but it is composed of three silicate layers of pseudo-hexagonal-symmetry. Therefore, its (001) cleavage plane has a two-dimensional pseudohexagonal symmetry. Mica is dehydrated at temperatures higher than 400°C and the structure of the cleavage surface changes remarkably when heated above that temperature [18]. The electron micrographs of Cu–Pc epitaxially grown on muscovite shown in Fig. 24 indicate differences of the surface before

Fig. 24 a, b. Epitaxial film of Cu–Pc grown on muscovite without (**a**) and with baking out (**b**) of the muscovite at 400°C before deposition of the Cu–Pc. Crystal growth is performed at 150°C in both cases; (by courtesy of Prof. M. Ashida)

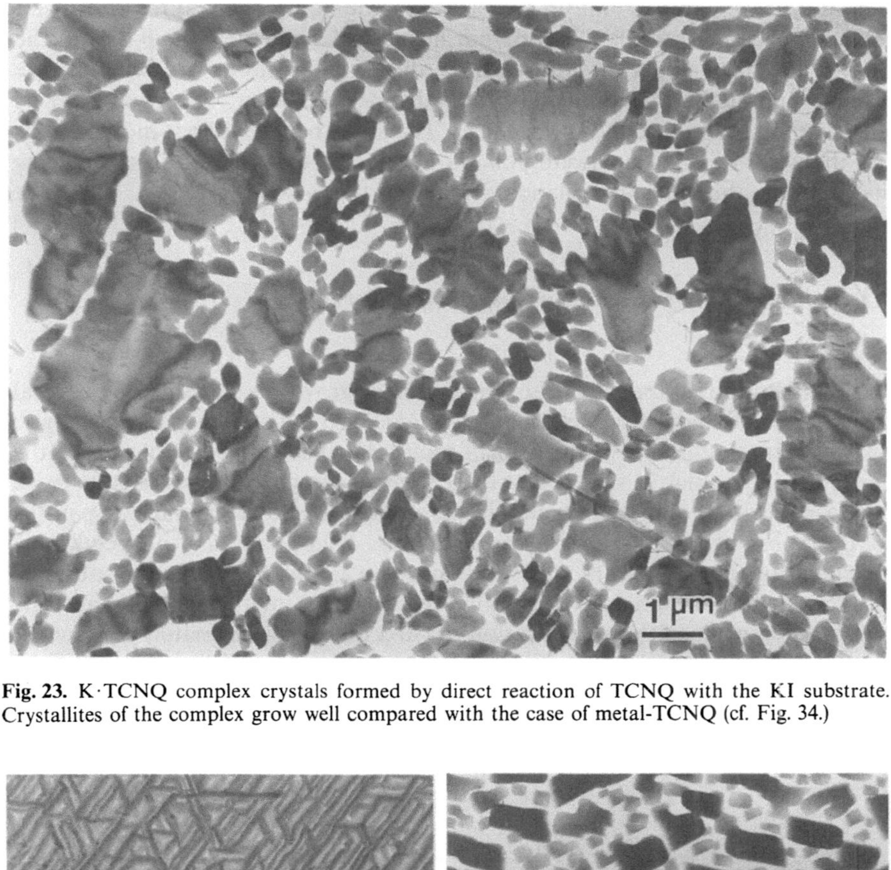

Fig. 23. K·TCNQ complex crystals formed by direct reaction of TCNQ with the KI substrate. Crystallites of the complex grow well compared with the case of metal-TCNQ (cf. Fig. 34.)

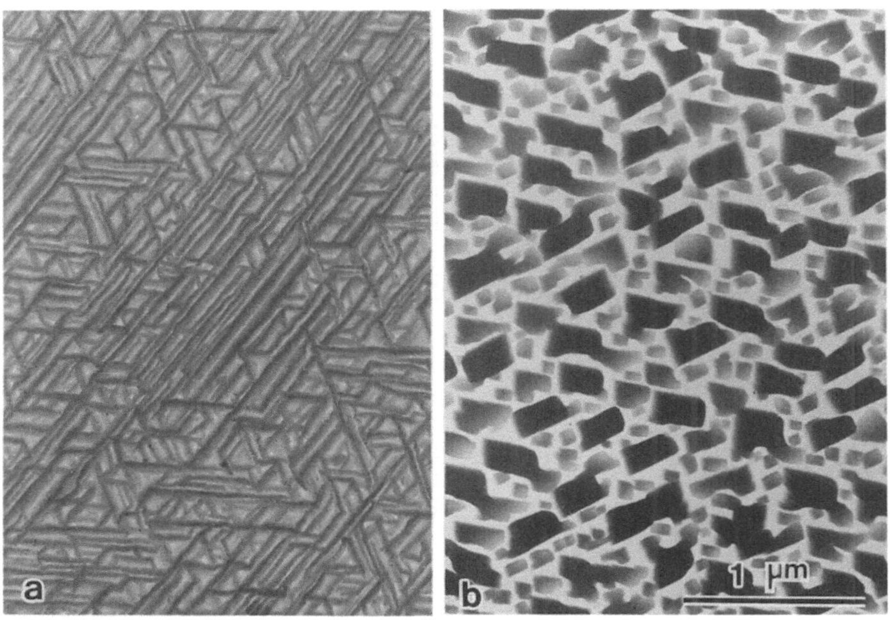

and after heating in vacuum up to 550°C for one hour. The texture of the film before the heating reveals the apparent hexagonal symmetry; the *b*-axis of the Cu–Pc, the fiber axis, elongates in three different directions on the substrate plane. After the heat treatment of the substrate, however, Pc grows with its *b*-axis almost normal to the surface [18]. The angle between the standing *b*-axis and the normal of the substrate plane is about 25° which is the same angle as that between the molecular plane normal and the *b*-axis.

3.3 Epitaxial Growth in One Single Orientation

As described in the above section, the epitaxial growth is governed by several factors and not all are clear yet. For the epitaxial growth of such planar aromatic hydrocarbon compounds, which form molecular columns by piling up of the molecules in parallel, it is clear that there are two principal modes of epitaxy. In one case the molecular column axes in the epitaxially grown crystals are parallel to the substrate crystal (we abbreviate this orientation as "*p*-orientation" hereafter) and in the other mode the columns are almost normal to the substrate face ("*n*-orientation"). In the *p*-orientation the molecules interact with the substrate only weekly, while in the *n*-orientation the first molecules intimately contact the substrate in that they orient their molecular planes parallel to the substrate

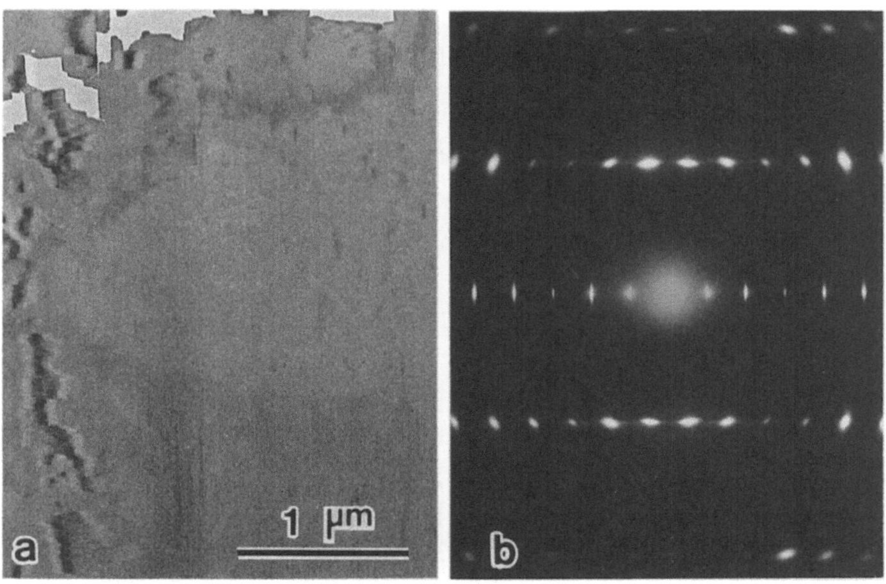

Fig. 25 a, b. Pt–Pc single crystalline film grown on muscovite (**a**), and its electron diffraction pattern (**b**); (by courtesy of Prof. M. Ashida)

surface. An example of the *p*-orientation is C_6Br_6 grown on alkali halides which were cleaved in air and not preheated (Fig. 19 a–c). This orientation appears usually on rock salt-type crystals and the long axes of the growing crystals are along [110] or [100] of the alkali halide. The (001) cleavage plane of these substrates has a high symmetry, and [110], and [$\bar{1}$10], as well as [100] and [010] are equivalent; therefore, epitaxial orientation along these equivalent directions occurs with the same probability. Accordingly it is very scarce that aromatic crystals grow on a (100) face of an alkali halide in only one direction. If one uses a muscovite plate as a substrate, one can obtain a single-crystalline film (Fig. 25a,b) of Pc or its divalent metal derivatives under appropriate conditions [18] while in usual cases the films are composed of fiber crystallites elongated in three directions reflecting the symmetry of the surface structure of the muscovite (001) face. Uyeda et al. have tried to interprete the formation mechanism of *single* orientation films on the basis of the Burger's vector defining the defect structure of mica [19].

In the case of perylene grown on a well backed out KCl (001) face the crystals assume a single orientation. Crystalline films in a single orientation can only be obtained when the *symmetry* of the substrate and of the molecular structure match. It is therefore difficult to find an appropriate substrate crystal which forces the nucleation process of a low symmetry organic material in a single orientation. The *lattice mismatch*, however, does not seem to be so important for organic epitaxy, owing to the large molecular and unit cell sizes as compared to typical lattice parameters of inorganic substrates.

3.4 Multiple Orientations of the Epitaxial Material

If the crystal structure of the epitaxially grown material has a high symmetry like in the case of gold on NaCl the epitaxial film grows in a single orientation. An *organic* epitaxial film, however, is usually composed of crystalline parts oriented in a number of directions. This is caused by the shape anisotopy of most molecules which leads to the low symmetry of most organic crystal forms, which are then not uniquely compatible with the high symmetry of the two-dimensional surface lattice of the substrate. In the case of a film grown on an alkali halide in *p*-orientation, for example C_6Br_6 on KCl, the *b*-axis aligns along the [110] direction of the substrate. Because the substrate [110] direction is symmetry-equivalent to the [1$\bar{1}$0] direction, the film is composed of crystals taking two or four perpendicular orientations [20]. For example, if it is preferable for the crystal *b*-axis to deviate with a definite angle ϕ ($\phi < 45°$), from the [110] and, therefore, the [1$\bar{1}$0] direction, the orientations with the angle $-\phi$ are also equivalent, and in such case at least four orientations are possible. When the $+b$ and $-b$ axis of the organic crystal considered are not equivalent, eight orientations can be distinguished. The cause responsible for the occurrence of the *p*-orientation is not clear but it may have to do with long-range forces acting between the substrate and the overgrowing crystal, because epitaxial growth in this orientation can occur in poor vacuum or in air or even at growth from solution, too.

On the contrary, the mechanism leading to the *n*-oriented films can be explained by the following principle [20]. It is found that this orientation only occurs when the substrate crystal surface is fairly clean; the organic molecule can therefore directly interact with the atoms or ions of the surface. A single molecule which reaches the crystalline surface is considered to settle at the most stable site by surface diffusion whenever the surface temperature is kept at a suitable value. Then the molecule, which takes a definite orientation at the site, acts as an epitaxial nucleus for the surface crystal. That is, the direction of the crystal axis of the epitaxial film relative to the substrate is determined by the orientation of the *first* fixed molecule which controls the orientation and position on the surface of other molecules coming close to it. In general the cohesive forces among the molecules are stronger than those acting between the molecules and the substrate atoms or ions. Therefore, the *molecular* fit to the atom or ion arrangement at the substrate surface is more important for the epitaxial growth orientation than a so-called lattice mismatch, i.e. the size and angle differences between the unit cell dimensions of the substrate and the epi-crystal. This is the most remarkable conclusion concerning the mechanism of epitaxial growth of organic crystals, in contrast to the case of metals and inorganic materials. For example, Cu–Pc crystals grown on KCl take four different orientations, all of which can be derived simply from one fixed orientation of a first molecule. Such multi-orientation of the crystals may occur whenever a number of stable positions for fixing a molecule are possible as a consequence of the low molecular symmetry and the

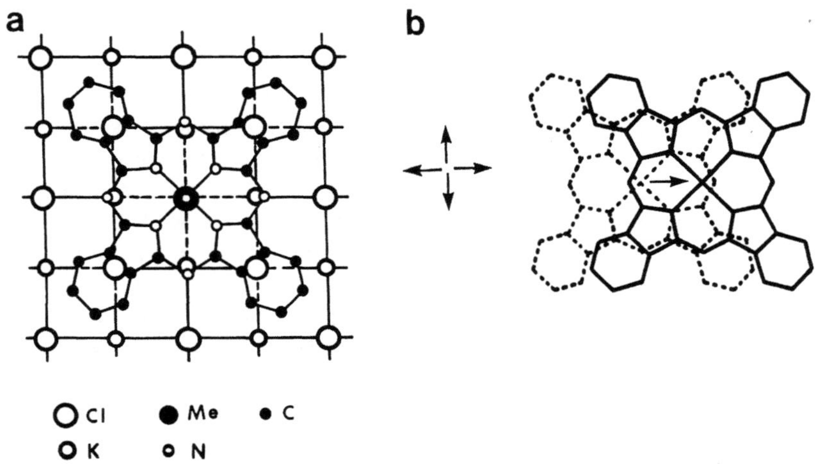

Fig. 26 a, b. Most plausible adsorption position of a Pc molecule on KCl (**a**). For the crystal growth from this adsorbed molecular nucleus, the molecule coming next can select four equivalent stacking sites from which four possible directions of the *b*-axis can result (**b**)

high symmetry of the structure of the substrate surface. These facts have been confirmed in several cases of planar molecules [21,22]. Cu–Pc and Pt–Pc are the examples studied most extensively. Both of these Pc molecules have been proved to assume a fixed orientation on the surface of KCl as shown in Fig. 26a. In this case, the Pc molecule adsorbed first takes a parallel orientation to the surface. With this adsorption geometry, the electronegative points of the molecule come to lie very close above the positive K^+ ions of KCl, which is apparently the most stable state for divalent metal-Pc's vapor deposited onto KCl or NaCl. The crystal *b*-axis of these materials coincides with the axis of their molecular columns. The columns are composed of molecules stacked in parallel but slipped relative to the underlying ones as shown in Fig. 26b. These molecules have D_{4h} symmetry and accordingly four equivalent directions are possible for the slip. As the result, four *b*-axes of the type shown in Fig. 26b may arise, emerging from the one Pc molecule adsorbed and fixed first. Thus, a distinct multiple orientation of crystal axes is fairly common with epitaxial films of organic materials.

3.5 Preferred Orientation

Even in absence of epitaxial interactions a so-called "preferred orientation" (a "texture") may occur. This can happen when the crystallites composing a film have a pronounced anisotropy (e.g. a tabular crystal habit). For example, if Cu–Pc needle crystals are dried on a glass plate from a suspension, they are fixed with the needle axes parallel to the glass surface. Since the *b*-axes coincide with the needle axes all *b*-axes lie in the plane of the glass plate, though with random azimuth. this state is also called a "preferred orientation" meaning that not all three, but only one or two crystallographic axes are fixed in space relative to the substrate. In this sense, a fiber structure may occur and the specified axis is called fiber axis. Electron diffraction patterns from a film with a fibrous structure or a film composed of crystallites in a preferred orientation occur in two different appearances, depending on the relationship between the directions of the fiber axis and the incident electron beam.

One appearance is observed when both directions are parallel to each other. In this case the electron diffraction pattern assumes a Debye–Scherrer ring pattern, but special rings are lacking in the pattern, namely those having the plane indices along the fiber axis as shown in Fig. 27. In this example (GeOPc) the fiber axis, the *c*-axis, is parallel to the incident beam and, accordingly, only the rings indexed (hk0) can be observed. The situation, represented in reciprocal space, is shown in Fig. 27b. Another essential difference between the Debye–Scherrer ring pattern and the fiber pattern exists in the possibility to form oblique patterns. When the film is tilted against the incident beam, the diffraction pattern shows a remarkable change and gives a so-called oblique fiber pattern. The oblique pattern and its schematic explanation are shown in Figs. 28 a, b. This type of

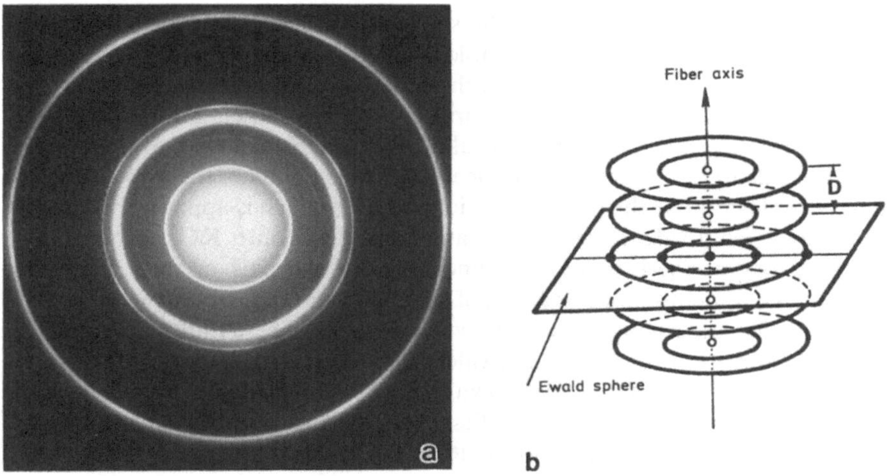

Fig. 27a. Electron diffraction pattern from poly-GeOPc grown on muscovite. The fiber axis is parallel to the incident beam. **b** illustrates the relation between the Ewald sphere and the reciprocal lattice

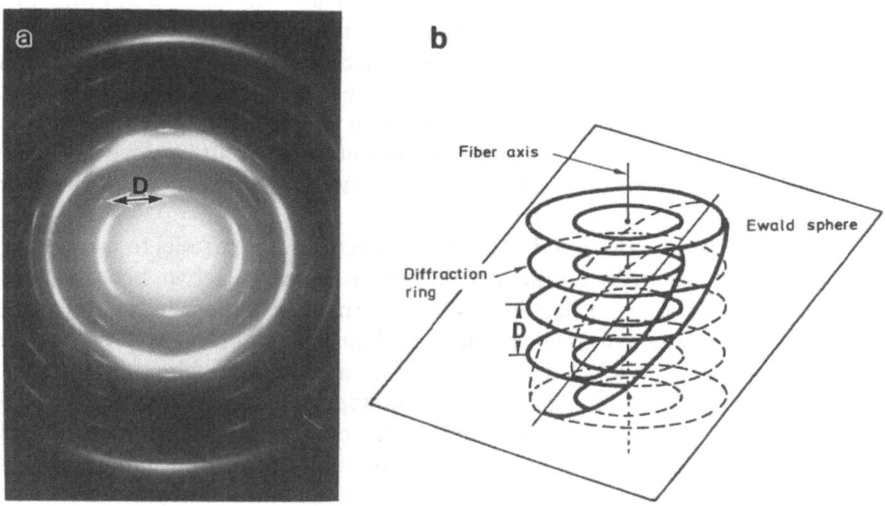

Fig. 28 a, b. Oblique fiber pattern obtained from the same sample as in Fig. 27a. The tilting angle is about 40°. The Ewald sphere makes an angle with the fiber axis and thus gives an elliptic pattern

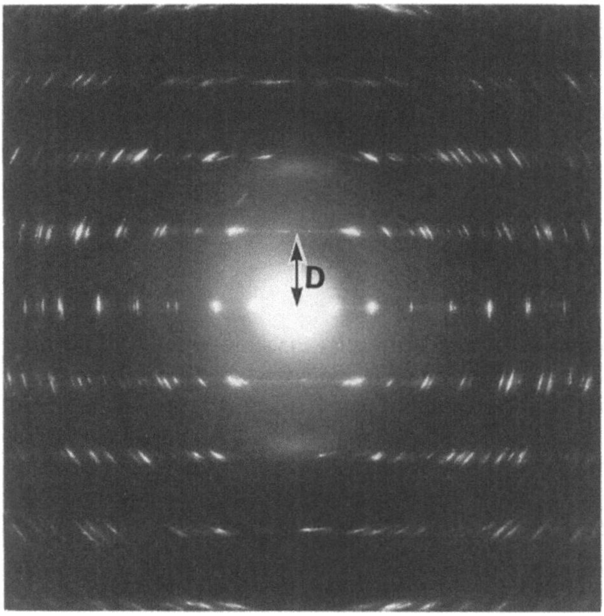

Fig. 29. A typical fiber or texture pattern obtained from $Si(OH)_2Pc$ grown on muscovite. The fiber period is indicated as D

pattern is useful for obtaining a three-dimensional information about the specimen and for the measurement of diffraction intensities from which one obtains a Fourier map of a potential distribution of the specimen as described in detail by Vainshtein [23]. On the other hand, a normal Debye–Scherrer ring pattern does not show any changes on tilting the sample.

The causes for the formation of such fibrous orientation are essential and also important for obtaining an oriented thin film. Geometric properties of the molecules and interactions between an individual molecule and the substrate are essential. A film of a condensed aromatic hydrocarbon or of a divalent metal derivative of Pc takes a preferred orientation even when these compounds are vacuum-deposited on a glass plate. These materials are hydrophobic while a glass surface is in general hydrophilic. Therefore, the molecules of these compounds place their molecular plane perpendicular to the glass surface when they deposit from the vapor. As a result, such a film takes a "preferred orientation", or "texture". This fact has been found by Inokuchi et al. by X-ray diffractometry [24] already in 1949. When the glass plate is pretreated by heating, its surface properties change; the crystallites can attain a different orientation on deposition, according to the molecular properties.

The second appearance is caused by a fiber axis which is oriented *in* the film plane and ordered. The diffraction pattern is resolved into single spots as shown in Fig. 29, but several series of diffraction spots are simultaneously found on their layer lines and on the equator. The regular interval between the layered lines is called a fiber period which is due to periodic order along the fiber axis. A tilting of the film about the fiber axis does not show any change while a rotation about the 0th layer line (equator) results in a change of the period and of the arrangements of spots on every layer.

3.6 Topotactic Reactions in the Solid State

One can obtain a film composed of well oriented crystallites of a molecular complex by a topotactic solid state reaction of a single crystal or an epitaxial film. The starting material is in most cases a thin film which reacts with a solid or with a vapor of another material to form, e.g., a charge transfer complex. TCNQ, for example, reacts with anthracene vapor or with an anthracene film vacuum-deposited onto a TCNQ crystal [25]. The crystal structure of the Anth:TCNQ complex that formes has been reported by Williams et al. [26] as follows:

TCNQ-Anthracene (1:1), a = 11.476 Å, b = 12.947 Å, c = 7.004 Å, $\beta = 105.4°$, Space group; C2/m

In this complex, the anthracene and TCNQ molecules form mixed stacks as schematically explained with two different projections in Fig. 30a,b. Electron micrographs of TCNQ formed on a KCl substrate and of its reaction product with anthracene vapor are shown in Fig. 31a,b. The reaction was performed at

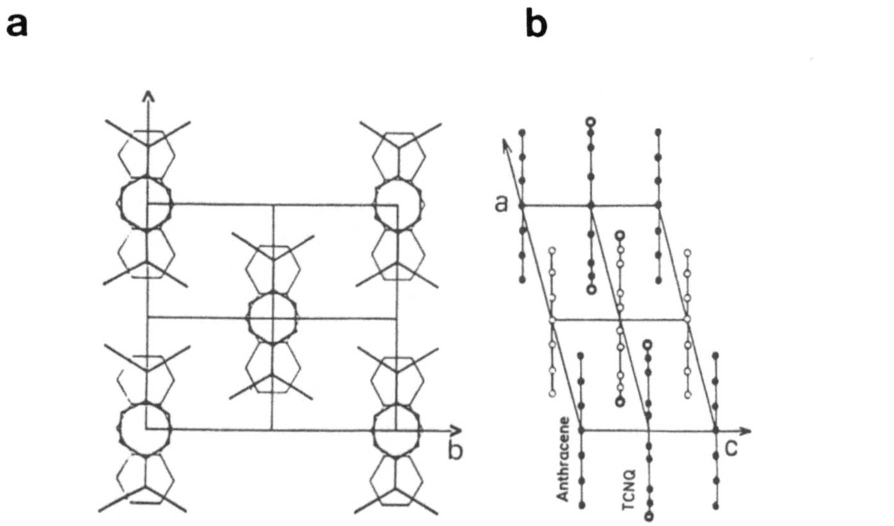

Fig. 30 a, b. Projection of the crystal structure of the anthracene·TCNQ complex along the *c*-axis (**a**) and along the *b*-axis (**b**)

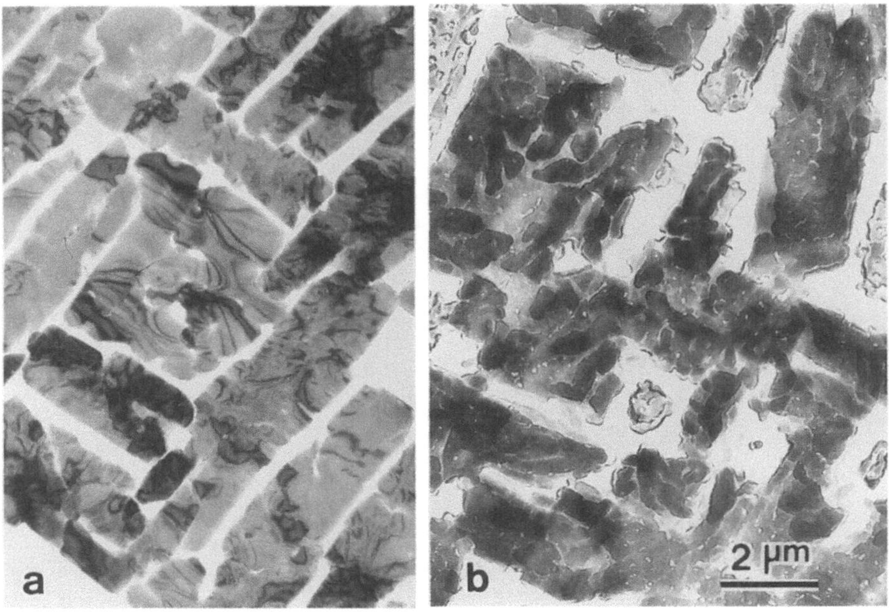

Fig. 31 a, b. Electron micrographs of the original epitaxial film of TCNQ (**a**) and itss reaction product with anthracene (**b**)

50°C. The color of the original film changed from yellow to brown. The electron diffraction at this stage shows that the individual crystals of TCNQ lying on their (010) habit faces partly remain unreacted while those of the complex lie on their (001) faces. The appearance of the complexed film changes and many small but thick crystallites are found to form. The reaction of TCNQ with the *vapor* of anthracene is fast but it proceeds slowly when an anthracene *film* is deposited onto TCNQ. The specimens at an intermediate stage of the reaction provide basic information on the orientational relation between the TCNQ matrix and the complex formed. The electron diffraction from the film at the reaction intermediate is exhibited in Fig. 32a together with a schematic drawing, Fig. 32b. Figure 33 reveals that the molecular arrangement of TCNQ in the complex is similar to that in the original crystal. Obviously, a close coincidence of the TCNQ positions before and after the reaction is found in the figure when the center of molecules are placed at the lattice origin as shown in Fig. 33. Thin lines in the figure indicate the lattice of the complex, whereas the bold ones represent that of the original TCNQ crystal; the molecules located at positions half way up or down in the unit cell are omitted for simplicity. Thus the transformation from the lattice of the matrix crystal to that of the complex can easily proceed by a slight displacement and a small rotation of the TCNQ molecules. In this case the anthracene, a relatively large molecule, must be intercalated by tearing off part of the TCNQ stack.

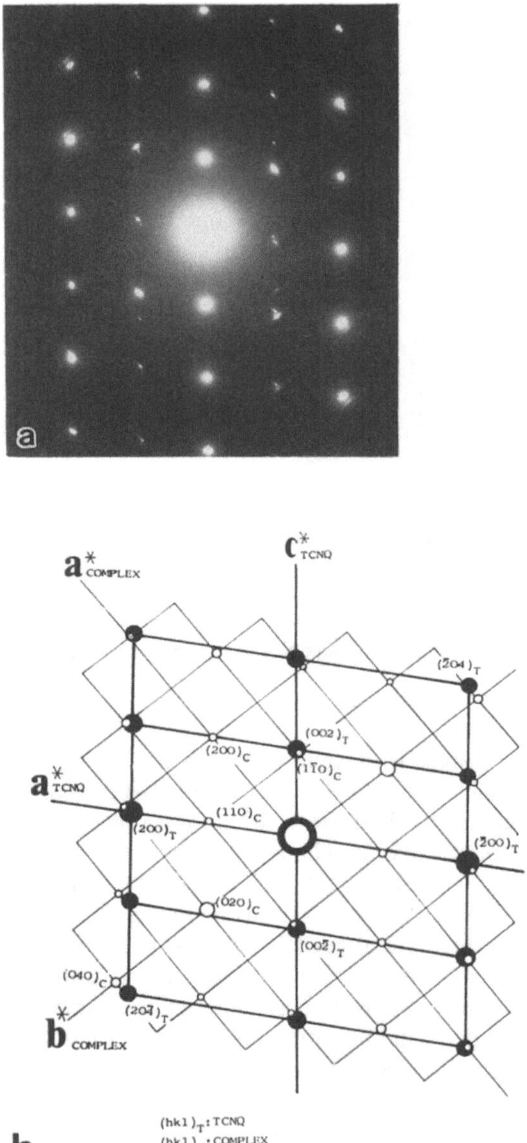

b

Fig. 32 a, b. Electron diffraction of the reaction intermediate (**a**) and cross section of the reciprocal lattices (**b**) showing the orientational relationship between the original TCNQ and the TCNQ-anthracene complex

In the next case, the reaction partner is a metal ion which is smaller than the molecules in the previous example which have anisotropic cohesive forces. When a metal is vacuum-deposited onto a surface of TCNQ which has been epitaxially grown on KCl, it reacts with the underlying TCNQ and forms a metal:TCNQ complex [27]. This reaction also occurs topotactically. Metals deposited on a

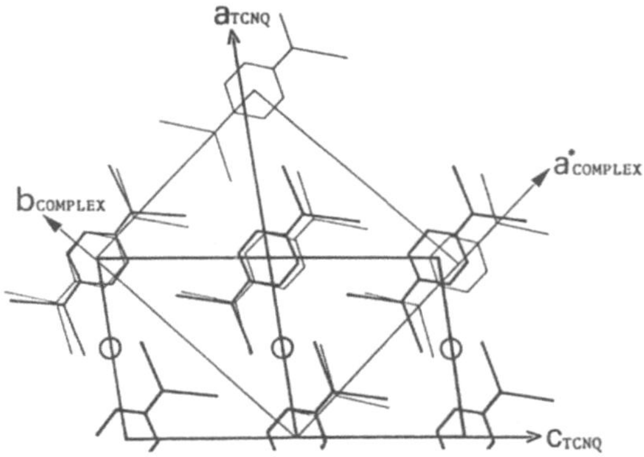

Fig. 33. Shift of TCNQ molecules on complex formation. The *bold lines* represent the two-dimensional unit cell and the molecules of original TCNQ crystal, while the *thin lines* are those of the complex crystal

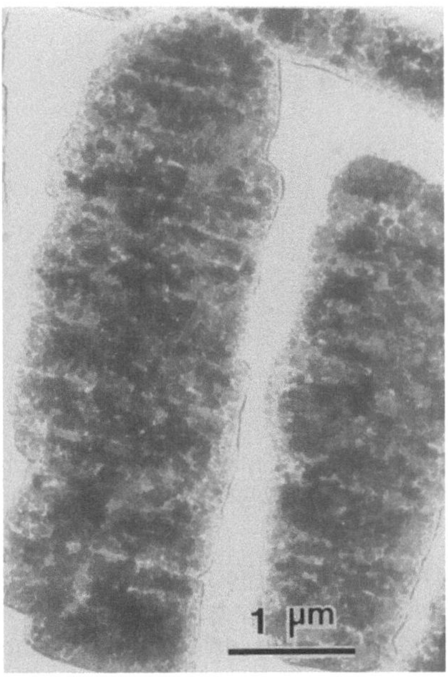

Fig. 34. Al·TCNQ complex, formed by direct reaction of TCNQ with an Al-film, vacuum-deposited on it. Many crystallites are produced in a mother crystal

TCNQ crystal as a thin film are ionized and diffuse into the TCNQ crystal. The product assumes a polycrystalline aggregate form as shown in Fig. 34 (the metal is Al in this case). However, the electron diffraction pattern of the aggregate reveals that every crystallite is oriented in such a way as if the aggregate formed a single crystal with a tetragonal lattice. A high-resolution electron micrograph of these aggregates is reproduced in Fig. 35 which shows a similar aspect as the images shown in Fig. 15. The image shows a square net pattern composed of obtuse zigzag lines and rod-like figures of 3 Å in width, both located in the middle

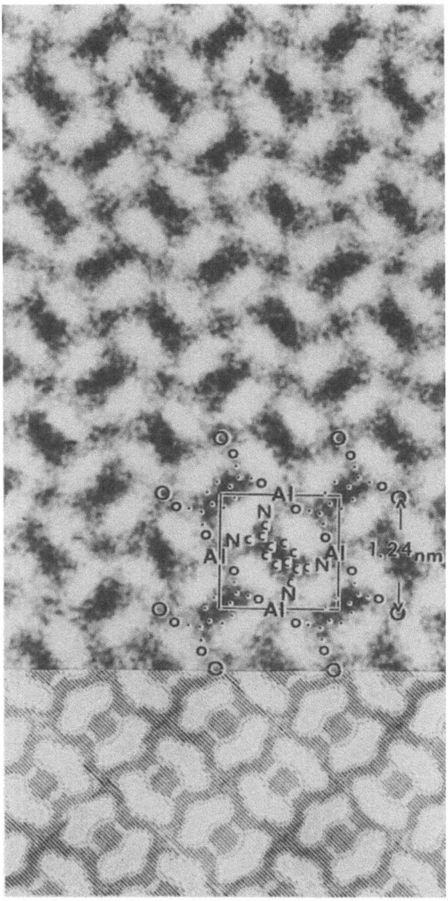

Fig. 35. High-resolution electron micrograph of the Al·TCNQ complex shown in Fig. 34

of the unit cell. The high-contrast dots at every knot of the net structure are assigned to be the aluminum ions. This structure is similar to that of the CT complex crystals formed between TCNQ and potassium or sodium. At the intermediate step of the reaction, which can be obtained by controlling the amount of metal deposited onto the TCNQ, both the product and the mother crystals co-exist. From the electron diffraction patterns of the specimen in such a composite state, the axial correlations between the crystals before and after the reaction can be derived. In this reaction the TCNQ molecules rotate by 45° clockwise and counterclockwise in the plane, with equivalent probabilities to form the complex crystal, as is shown in Fig. 36.

An "epitactic reaction" is also a useful method for obtaining well-oriented films. "Epitactic reaction" means that a reaction product produced on a single crystal surface grows with a definite orientation as an epitaxial film. For example, TCNQ, vacuum-deposited onto an epitaxial thin film of silver, reacts and the

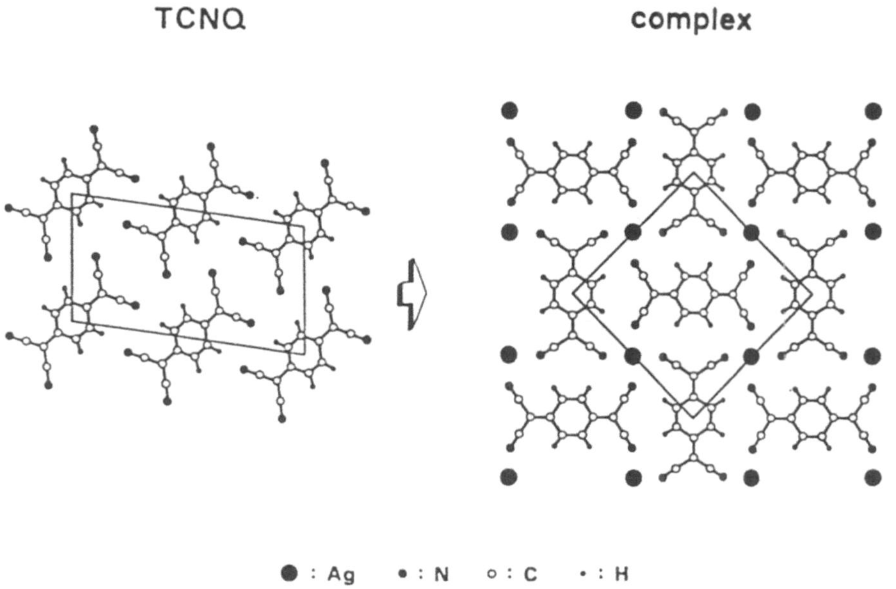

Fig. 36. Relationship between the molecular orientations of TCNQ before and after the reaction with a metal

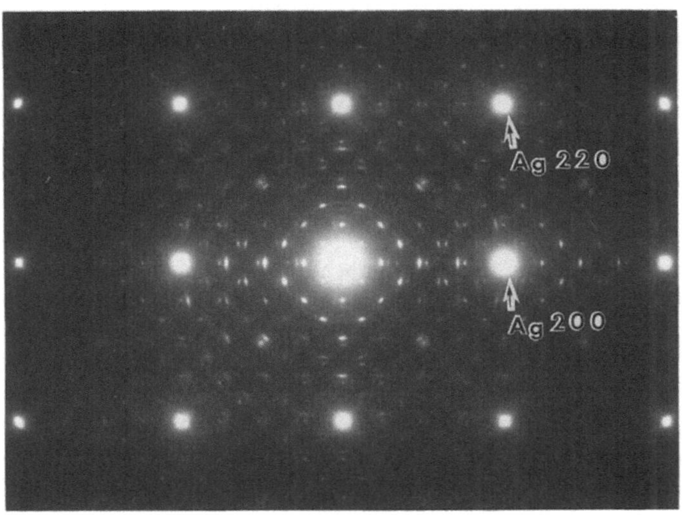

Fig. 37. High-resolution electron diffraction pattern from the Ag·TCNQ complex formed on an epitaxial Ag thin-film in (001) orientation. The pattern is similar to those shown in Fig. 15 but the unit cell dimensions differ

Ag:TCNQ CT complex grows in the form of a thin film which can assume a single crystalline structure. This crystal has the same appearance and lattice symmetry as in the case of a silver film deposited on TCNQ and reacted, but the lattice constants differ from the former case. The electron diffraction pattern of such a film is reproduced in Fig. 37, in which the single net pattern of the silver substrate is superimposed. The formation mechanism of this complex is not yet clear.

4 Observation of the Molecular Orientation at Crystal Structure Defects

If disorder exists in a crystal, an anomalous diffraction pattern can be observed because the diffracted beam deviates slightly from the Bragg condition. By using this fact and the diffraction contrast appearing in the electron micrographs, Amelinckx et al. and other investigators [28] have accumulated many data on defect structures in metals and other simple inorganic materials. However, only few researchers applied this method to organic crystals [29]. Organic molecules are in general anisotropic in their shape as well as in their cohesive forces. Therefore, from the diffraction contrast and its anomaly one cannot unambiguously determine the structure of the defect, and especially not the orientation and the real positions of the molecules, except for very few cases. On the contrary, high-resolution images based on phase contrast can reveal such structural situation of the defect site. In this section, orientations of molecules which deviate from the regular site of the crystal will be shown with the intention to discuss the structure from the viewpoint of the anisotropic features of these organic molecules.

4.1 Dislocations in Organic Thin Films

Among many other crystal defects, the edge dislocation is especially well known among electron microscopists because it is the first example of a defect which has been visually detected as a lattice image by high-resolution electron microscopy. In 1956 Menter et al. [1] succeeded for the first time in observing lattice images and proved the existence of the edge dislocation in Pt–Pc crystals. The lattice spacing they observed was 11.9 Å which corresponds to the $(20\bar{1})$ lattice planes. After that a keen competition in constructing higher-resolution electron microscopes has set in and the results were demonstrated by showing observations of narrower lattice spacings. At present one can observe lattice spacings as narrow as 0.5 Å. However, the real resolution of an electron microscope should be characterized by a theoretical point-to-point resolution, defined by the electron

Fig. 38. Dislocations in Pt–Pc. Strain of the lattice planes is unsymmetrical

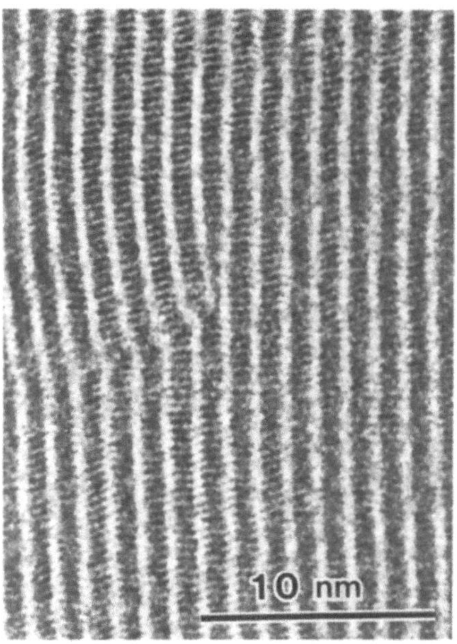

Fig. 39. Enlarged image of a part of Fig. 38. The molecular stacking in each wide lattice line differs locally but at the dislocation core the orientation is unclear

wavelength and the spherical aberration constant of the objective lens, and should be distinguished from the *lattice resolution*, though the lattice resolution can give an information on mechanical and electrical stability of the microscope. Figure 38 shows the lattice image of a Pt–Pc epitaxial film grown on mica [30]. The displayed area contains many edge dislocations. The dislocation lines are perpendicular to the plane of the paper. The high-resolution electron micrograph can even reveal the molecular stacking in each lattice line as shown in the magnified Fig. 39. The molecular columns of Pt–Pc are projected on a plane which includes the column axis. The molecular image at the dislocation core is not so distinct in the figure. However, an illustration shown in Fig. 40 well explains the situation.

The softness or plasticity of a metal is, in general, caused by the presence of slip planes. Applied stress is relaxed by the movement of dislocations in which metal atoms are displaced along a slip plane that includes the dislocation line, and the dislocation can travel from one place to another as if the top half of the crystal was displaced relative to the bottom half. On the contrary, a molecular crystal of an aromatic compound is not always sufficiently soft and is therefore easily broken by a shear stress. This is partly due to the weak cohesion forces of the molecular crystal, which are mainly of van der Waals nature, but partly also due to an immobility of the dislocation. The anisotropic nature of the cohesion forces as well as of the molecular shape disturbs the movement of the dislocation.

The structure of Pt–Pc has been reinvestigated by X-ray diffraction by Brown [31] and shown to belong to the space group C2/n. The epitaxial films of the Pc compounds grown on muscovite are isomorphic with the α-form of Pt–Pc reported by Ashida et al. [17] on the basis of the high-resolution electron diffraction method. However, a recent investigation by high-resolution microscopy shows that the epitaxial film is composed of many polymorphic crystallites in which the arrangement of the molecular columns varies, though the structure of each molecular column is identical with that in the α-form of the Pt–Pc crystal. The Pc molecules stand in a line to form molecular columns in which the angle between the normal to the molecular plane and the column axis is 25.6°. The distance between the mean planes of two adjacent molecules is 3.45 Å. This column structure is maintained in the vacuum-deposited films of divalent Pc derivatives, but the arrangements of such columns are locally different, even

Fig. 40. Schematic representation of the expected molecular orientation at an assumed dislocation. At the lower half of the crystal, the arrangement of the column requires –AA– or –BB– stacking instead of –ABAB–

Fig. 41. It is schematically explained in thhe figure that an unnatural molecular packing would be needed for the dislocation to move

within each individual crystallite. As a consequence of this diversity of the arrangements, polymorphism and stacking faults are easily produced. When the projection of the column on the plane including the column axis is considered, two fundamental situations are possible. In one situation an image is observed in which the molecular plane makes an angle of about 26° with the column axis (A-column), and in the other is its mirror image appears where the angle is − 26° (B-column), as illustrated in Fig. 40. In most cases the columns A and B stack alternately in a plane as shown in the upper part of Fig. 40. At an edge dislocation, as seen in the micrograph Fig. 39, the situation in the --ABAB-- stacking changes on the half crystal (lower part of Fig. 40). That is, a lack or an insertion of a half-plane results in --BAAB-- (or --ABBA--) stacking (Fig. 40). In order for the fault to move like in the case of metals, the A (or B) column in the upper half must join with the adjacent B (or A) column of the lower half crystal (Fig. 41). Thus a slip along the molecular plane may easily occur in a column, but the slip accompanying the movement of the dislocation is required to take place normal to the column axis and thus is hardly expected to occur.

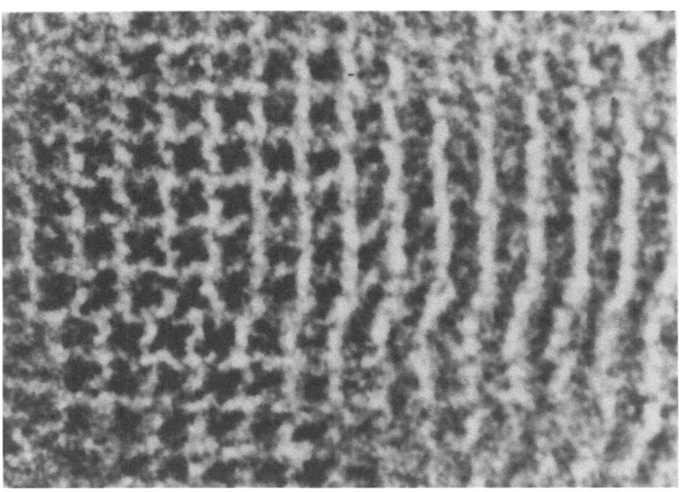

Fig. 42. An edge dislocation which appeared in the *a*–*c* plane of Zn–Pc. The orientation of the molecules is reversed in the upper half of the crystal with respect to the lower one

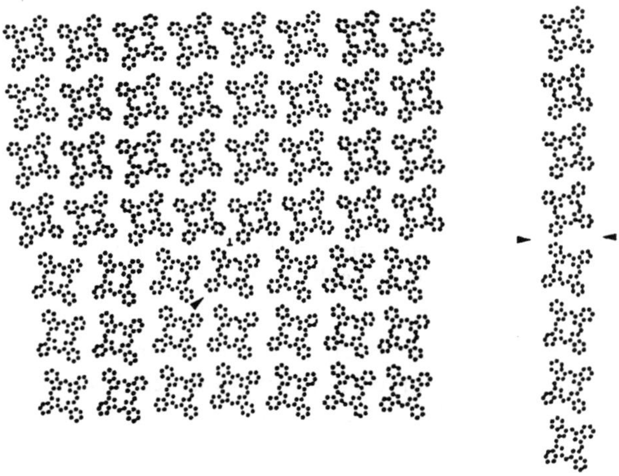

Fig. 43. Schematic presentation of the dislocation shown in Fig. 42. In order to relax the lattice strain caused by the dislocation, molecules shift toward the core, but the shift requires a molecular rotation as observed in Fig. 56. Apart from the core, such molecular rotation disturbs the close-packing of molecules and results in the production of a disordered structure as observed at the right part of the depicted lattice image

We can hardly find an edge dislocation whose dislocation line runs parallel to the molecular column, and the micrograph shown in Fig. 42 is a very rare case which occurred in a small and thin crystallite [32]. The difficulty for this dislocation to appear can be explained by the anisotropy of the cohesive forces of the molecules and is illustrated in Fig. 43.

4.2 Stacking Faults

Stacking faults are known as planar defects and commonly observed in metals. In the case of organic crystals, the faults are caused by the anisotropic nature of the molecules and of the intermolecular cohesive forces. Contrary to an inorganic crystal, a certain different kind of stacking fault, a line defect, frequently occurs in organic crystals, which is caused by the weakness of thc interactions between the molecular columns. That is, the individual molecular columns behave as if they were long-chain molecules. When this type of "quasimolecule" stacks in a false way in the crystal, the defect may be called a linear stacking fault. The cohesive force of planar molecules is stronger in the direction of the plane normal than in a parallel direction. Accordingly, displacements along the column axis are frequently observed as stacking faults. Hexabromobenzene is a typical planar molecule which also forms molecular columns. When the crystal is vapor-deposited onto a clean (001) face of molybdenite (MoS_2), the molecules are tightly adsorbed on the substrate as illustrated in Fig. 22. Growth of the molecular

Fig. 44. Stacking faults in a C_6Br_6 film introduced by the interaction between the molecules and the substrate surface

columns may occur from every adsorbed molecule whose plane normal makes an angle of 23° with respect to the column axis. The crystal is composed of such columns but the lower ends of all columns are fixed to the substrate. Therefore, the arrangement of the columns relative to the neighboring ones does not occur in the ordinary way, but the columns display a slippage along the column axis. An electron micrograph shown in Fig. 44 represents such stacking faults along the *b*-axis. The pair of electron diffraction patterns shown in Fig. 10a,b has been obtained from C_6Br_6 crystals on MoS_2. When the incident electron beam is normal to the substrate, many sharp streaks running along the *a*-axis can be observed on every spot, whereas the streaks on the diffraction spots disappear, as shown in Fig. 44b, when the specimen is tilted about 23° relative to the incident beam so that the beam direction coincides with the column axis. This is because the displacement along the column cannot be seen from the direction of the column axis. This case of stacking fault originates from the strong interaction between the substrate and the first molecules deposited onto it.

The next case is a stacking fault caused by a misorientation of a molecular column during growth. Zn–Pc films, vacuum-deposited onto a thin carbon film, contain many crystal defects [32], one of which is shown in Fig. 45, where the molecules are projected along their planes and observed as stripes. The molecules are tilted approximately 26° relative to the column axis and columns with

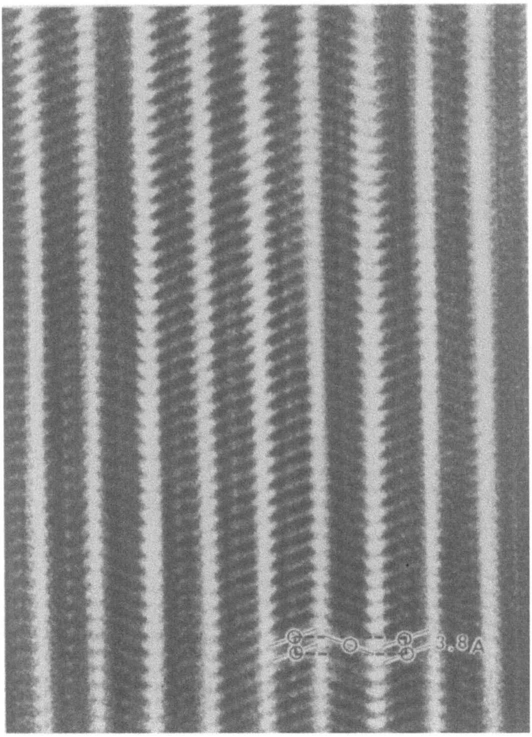

Fig. 45. Stacking fault or misorientation of molecular columns found in Zn-Pc. This fault is frequently observed in the vacuum-deposited films of divalent metal Pc's

a tilting angle of $+26°$ or $-26°$ usually stack alternately in the crystals. However, sometimes the alternate arrangement is disordered as shown in the micrograph. A successive occurrence of such disorder has been found in some films looking like as if they contained polymorphic forms. This fact suggests that the cohesive energy of the molecular columns does not differ appreciably between the stacking patterns of the --ABAB--and that of the--AAA--or --BBB--type, a fact which leads to the formation of polymorphs as well as dislocations.

Another kind of defect is that introduced in a thin crystal of the electronically conductive molecular complex of TTF:TCNQ by non-stoichiometry of the complex [12]. This case is also a stacking fault, but a line defect. The crystal of TTF:TCNQ is composed of molecular columns of TTF and of TCNQ and forms a "segregated stack" columnar structure. Each kind of these columns is again alternately stacked along the *a*-axis, with the column axis parallel to the *c*-axis. When the crystal is deposited onto a (001) face of KCl from the vapor phase, it grows epitaxially directing its *b*-axis along the two equivalent [110] directions of KCl. The electron diffraction pattern of the film of TTF:TCNQ (Fig. 13a) shows sharp streaks on every layer line including the equator. The streaks run along the layer lines and perpendicular to the *b*-axis. This fact indicates that stacking faults

Fig. 46. Stacking fault of a poly-GeOPc epitaxial film

occur along the *b*-axis. Usually an elongation of a reciprocal lattice point due to a stacking fault occurs in a direction perpendicular to the fault plane, therefore, the streak in the diffraction pattern is observable only when the faulted plane is parallel to the incident beam. If the fault plane is tilted against the incident electron beam, the streaks will disappear. In the present case, however, the streaks do not disappear when the crystal is rotated (Fig. 13b) around the *b*-axis. This indicates that the diffraction spots spread over in two-dimensions, i.e. a reciprocal wall is formed which corresponds to a linear defect in the real crystal.

The next example is a simple plane defect, a stacking fault, found in the epitaxial film of poly-GeOPc as shown in the electron micrograph of Fig. 46 [21]. The GeO–Pc forms a polymer chain instead of the molecular column in that it is linked together through—Ge—O—Ge—bonds, as illustrated in Fig. 47(a). In the chain, Pc ligands stack in parallel but stagger alternately with the angle of 37° relative to the adjacent ligands (Fig. 47b). The molecular images of Fig. 46 are the projection of the chains along the chain axis and show the same feature as illustrated in Fig. 47b. The electron diffraction (Fig. 48) of this crystal has sharp streaks on the diffraction spots which is evidently due to the stacking fault shown in the micrograph. The reason why the streaks run along both, the *a*- and *b*-axis, lies in the tetragonal symmetry of the crystal.

The characteristic features of the defects are:

1) The stacking faults occur always parallel to the (100) or (010) planes.
2) The displacement along the fault planes is one half of the unit cell vector along the *a*- or *b*-axis.
3) The spacing between two successive molecular layers, including the fault plane, maintains its original value of the unperturbed lattice.

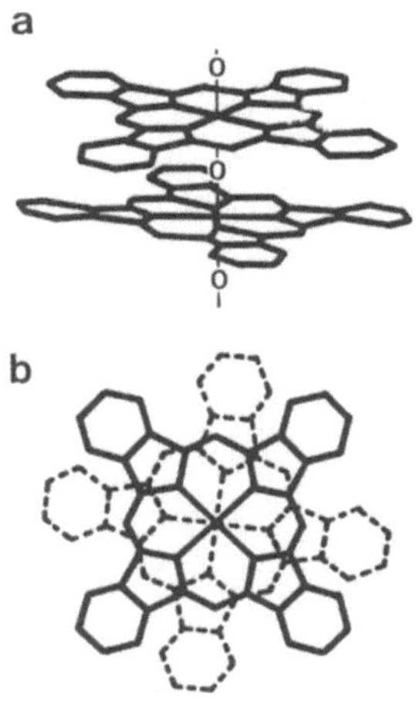

Fig. 47 a. A polyphthalocyanine chain formed by a linkage of the central germanium atoms by oxygen and a rotation of the Pc ligands about the chain axis. **b** Projection of poly-GeOPc along the chain axis. Staggering angle is 37°

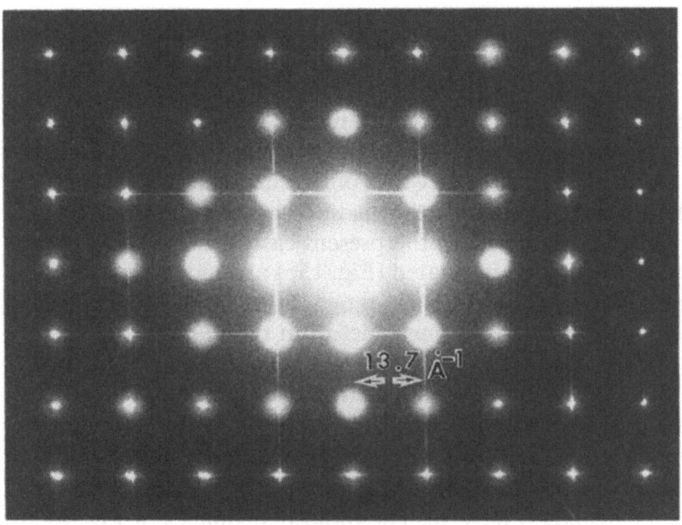

Fig. 48. Selected area electron diffraction from a crystallite of poly-GeOPc epitaxially grown on KCl. The streaks appear on the layer lines or on lateral lines of odd order along the lines

The molecular arrangement in the bottom layer of the crystal is shown in Fig. 49 as a space-filling model where each Pc molecule is closely packed keeping the van der Waals radii of carbon (1.7 Å) and hydrogen (1.0 Å) distances. Evidently it is difficult for the crystal to shift along the (100) or (010) plane by half a period without changing the molecular separation, because otherwise an overlap of compactly packed molecules would take place as hatched in Fig. 50. A similar situation has been observed in the case of Zn–Pc (Fig. 56) whose molecular orientation at one side of the boundary changes with respect to that

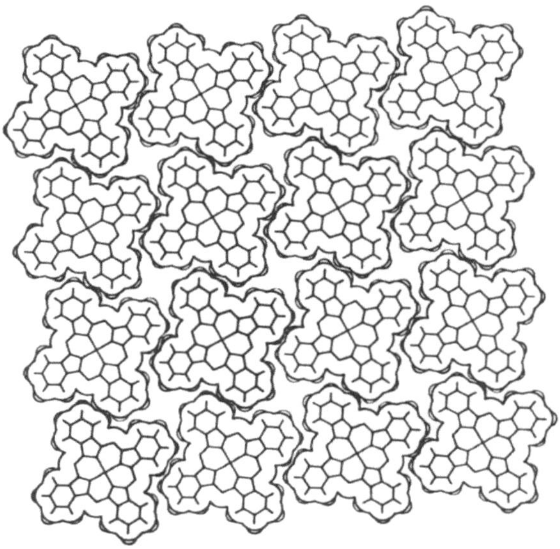

Fig. 49. Molecular arrangement in one layer of poly-GeOPc. In the layers above and below molecules are staggered around the molecular center by an angle of 37°

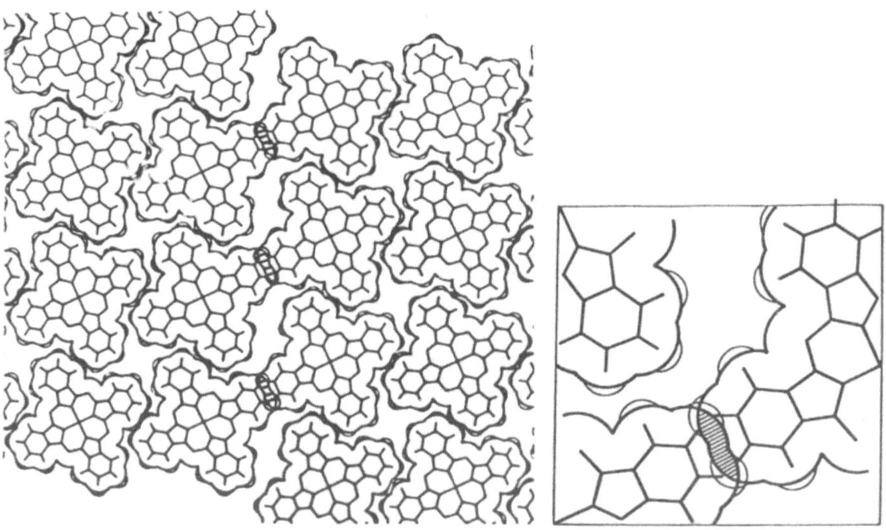

Fig. 50. The displacement of a half crystal along the *a*- or *b*-axis; without a change in spacing a facing of the fault plane is impossible due to steric hindrance (if the displacement occurs in the plane of the paper)

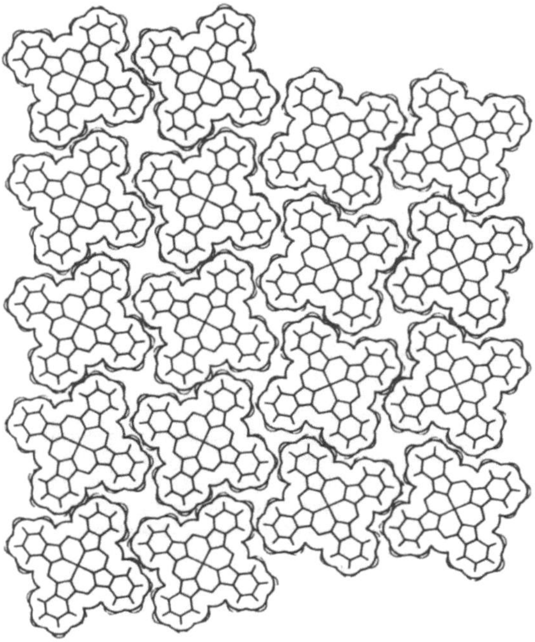

Fig. 51. The stacking fault produced by the displacement of 1/2(a + c) or 1/2(b + c) results in a close packing. (the c-axis is perpendicular to the plane of the paper)

at the other side. On the basis of this example the following structure model is derived which avoids the problem of steric hindrance. Suppose that at one side of the fault the molecules in the same layer rotate by 37° relative to the other side; as a consequence, the molecular packing becomes compact and no steric hindrance remains, as shown in Fig. 51. The arrangement of molecules in the right side of the figure is the same as that of the second layer in the left part of the figure. Consequently, the second layer in the right part requires the same arrangement as that in the first layer in the left. In other words, the fault plane assumes the property of a glide plane along the *a*-axis (or *b*-axis) and these stacking faults can be represented with a displacement vector of $\frac{1}{2}(a + c)$ or $\frac{1}{2}(b + c)$. When this stacking fault occurs in an alternating succession, the corresponding crystal part can be considered a polymorph, as reported in Ref. [33].

The electron micrograph shown in Fig. 52 exhibits the existence of fragmental space in the crystal due to the introduction of the stacking fault of this kind which ends in the middle of the crystal. The space is only one half of the unit cell dimension in width and a full molecular column can not be admitted by this space. However, the image contrast shows that some inclusions exist in this "hollow" space. To analyze and to reveal the identity and role of such inclusions are very important from both, a physical and a crystallographic point of view; but it must be left to future investigations, based on much higher electron microscopic resolution.

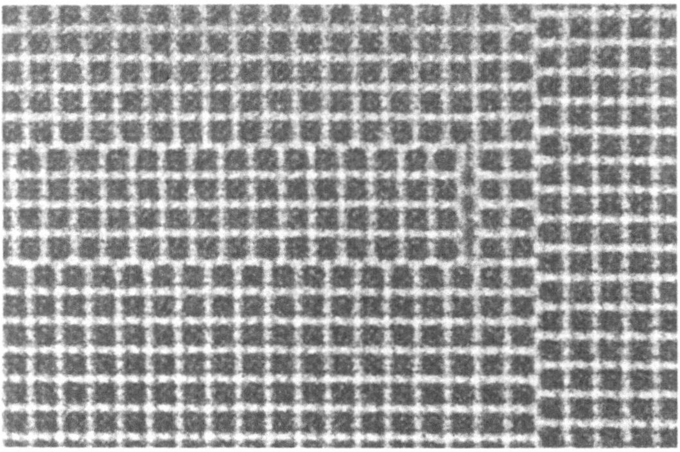

Fig. 52. An example of fragmental space produced by two stacking faults ending in the crystal. The space is filled by unknown molecules; such impurities usually play an important role in the properties of the crystal through the local changes in the electronic interactions they induce

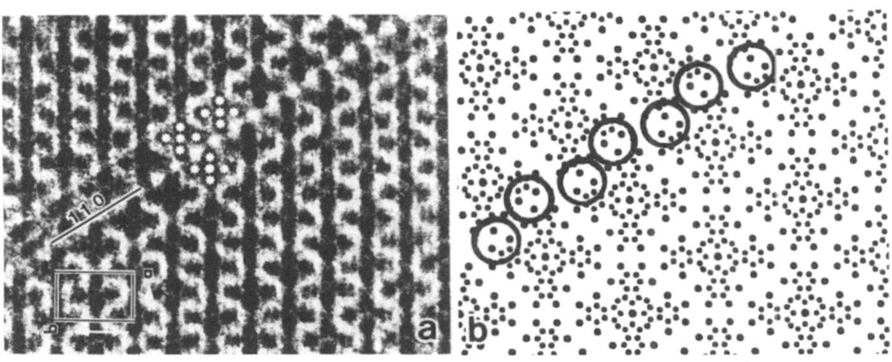

Fig. 53 a. Stacking fault found in the *a–b* plane of a *c*-centered crystal. The sample is the $Cl_{16}CuPc$ epitaxial film shown in Figs. 4 and 5. The displacement is about $\frac{1}{3}$ [110]. **b** Model of the fault. Note that the Cl–Cl interactions at the fault plane differ little from the ordinary part of the crystal. The interactions determine the packing energy along the plane

When a crystal belongs to the space group C2/c, a stacking fault occurs on the (110) plane along the [110]-axis. Smith et al. [34] have found this case in an epitaxial film of chlorinated Cu–Pc. The displacement of the lattice is $\frac{1}{3}$(a + b) as shown in the micrograph (Fig. 53a) and illustrated in Fig. 53b. The cohesive forces of an organic solid may be expressed by the summation of atom–atom potentials. In the crystal of this material, chlorine-to-chlorine interactions seem important for the molecular columns to cohere. As illustrated in the Fig. 53b such interactions among the chlorine atoms appear to be almost the same at an ordinary and at a fault plane. If one knows the atom coordinates along the projecting axis of the crystal, one can calculate the stacking fault energy in this case.

4.3 Grain Boundaries

Physical properties of grain boundaries depend on their structure. However, definite correspondence between the properties and the structure is not yet clear. The atomic arrangement at the real grain boundary was essentially unknown until recently; modern high-resolution electron microscopy has made it possible to observe, in principle, the structure of the grain boundary at atomic resolution. However, as to organic crystals, very little has been reported on this subject, because of the irradiation damage problem arising with organic material. Pc derivatives are organic crystals exceptionally resistive against electron irradiation and therefore have been studied by electron microscopy by several investigators. A twin boundary is easily observed because of its regular structure. The micrograph shown in Fig. 54a displays a crystal of Zn–Pc grown on amorphous carbon [32]. This crystal form is the α_{III}-type shown in the same figure (b). Well-defined images of rows of molecular columns and of unresolved molecules running parallel to the *a*-axis are arranged in an alternating sequence.

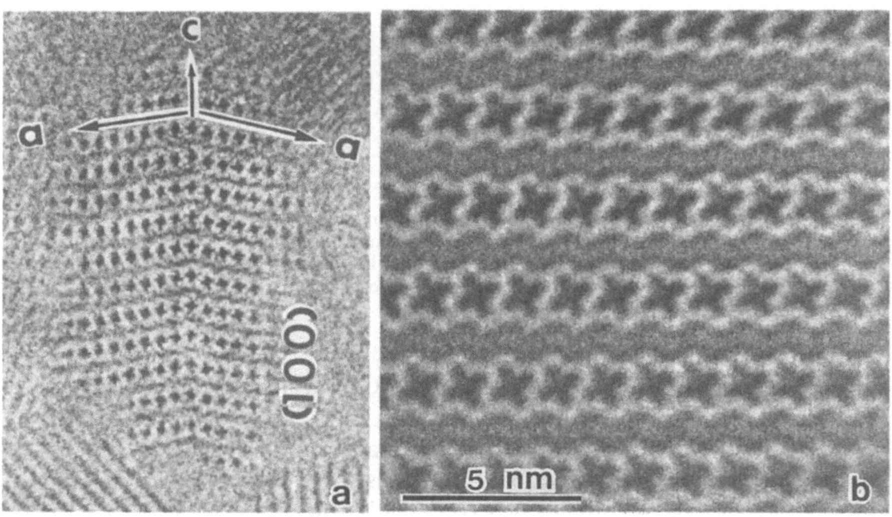

Fig. 54 a. Molecular arrangement at a twin boundary in Zn–Pc; **b** defect-free structure of a newly found Zn–Pc polymorph vacuum-deposited on an amorphous carbon film

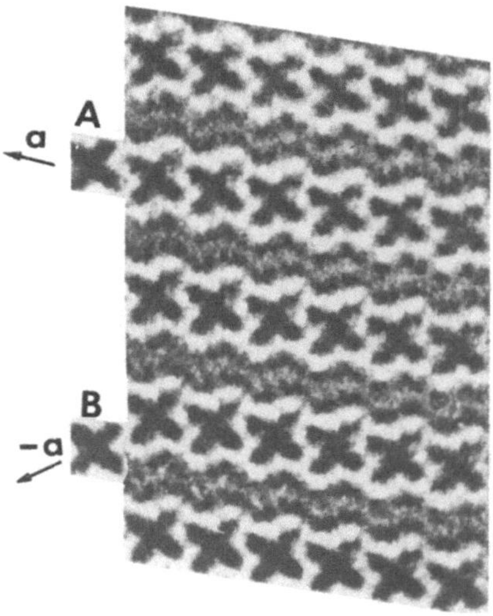

Fig. 55. Explanation of twin formation. Due to the molecular symmetry, the sites A and B may be equivalent energetically. A produces the twin and B continues the normal lattice

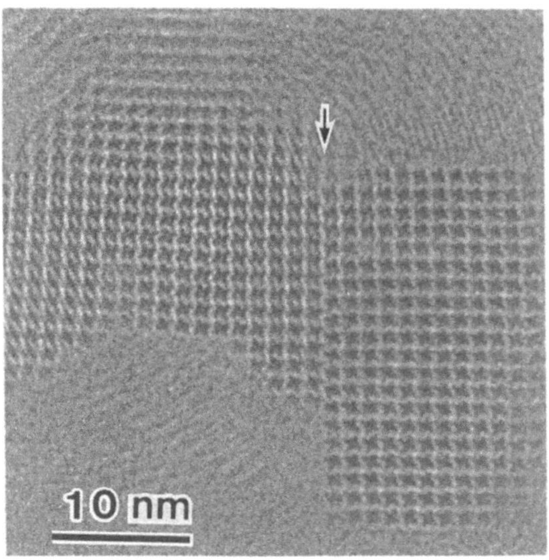

Fig. 56. Grain boundary of Zn–Pc which can also be considered to be produced by a false stacking of the molecules (see main text)

The twin boundary is the (100) plane; the *c*-axis is common to the crystals at both sides of the boundary. Evidently the *b*-axes, which are perpendicular to the plane of the paper, are "reversed" at the left and right parts. A possible formation mechanism of the Zn–Pc twins can be proposed on the basis of the anisotropy and symmetry of the molecular shape. Assuming the (100) plane as a growth front and taking the D_{4h} symmetry of the molecules into consideration, two possible sites exist for a new coming molecule, see Fig. 55. When the addition of the molecule occurs with the same orientation as that of the others in the crystal site A, the normal crystal growth continues; but when the molecule is attached in the orientation B, twinning occurs, starting from this lattice point. The cohesive energy between two molecules of Zn–Pc which has D_{4h} symmetry is considered to be equivalent in the two situations A and B shown in Fig. 55, where for simplicity only two interacting molecules are drawn.

A grain boundary produced by this cause can be seen in Fig. 56 which was also found in a Zn–Pc film formed on a carbon film. It can be observed in Fig. 56 that the molecular orientation on the left side of the boundary, indicated by the arrow is a (glide) mirror image of that in the right part. The molecules on the boundary take an orientation which is different from that of the molecules on both sides. The lattice displacement along the boundary can be seen in the image. This displacement and the inversion of the molecular orientation are considered to be in close correlation with the molecular shape and packing energy. If only a slip occurs in the plane, as in Fig. 56, and the displacement distance is a half period along the slip plane, the boundary becomes very unstable, owing to the steric hindrance on the slip plane, as shown in Fig. 57a. Such a boundary probably

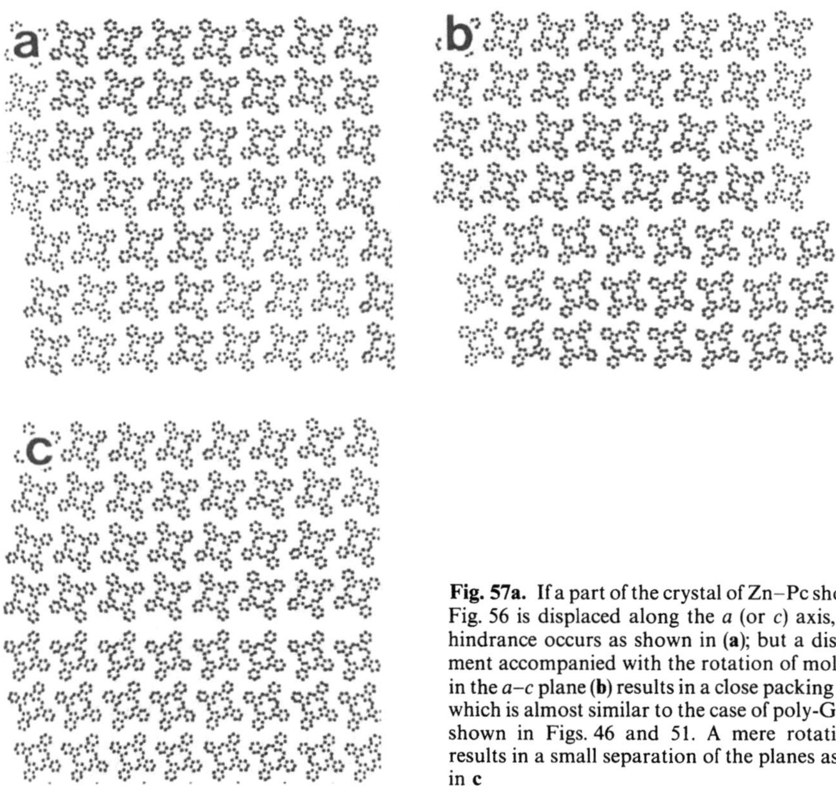

Fig. 57a. If a part of the crystal of Zn–Pc shown in Fig. 56 is displaced along the *a* (or *c*) axis, steric hindrance occurs as shown in (**a**); but a displacement accompanied with the rotation of molecules in the *a–c* plane (**b**) results in a close packing again, which is almost similar to the case of poly-GeOPc, shown in Figs. 46 and 51. A mere rotation (**c**) results in a small separation of the planes as show in **c**

cannot exist. However, when the slip is accompanied by a rotation of molecules in the plane around each molecular center in the lower half of the crystal, such steric hindrance can be avoided, as shown in Fig. 57b, which reflects the same situation as the real images shown in Fig. 56. If two crystals with inverse molecular orientations make a boundary without any lattice displacement along this boundary, a similar unstable configuration may be fomed at the boundary (Fig. 57c). In this case a displacement of a half period along the boundary is necessary to diminish the repulsive force and to increase the cohesive force between two adjacent molecules on each side of the boundary. Thus the parallel displacement of half the crystal, i.e. slip, and the rotational displacement of molecules should occur together in order to make the grain boundary stable, as shown in the Figs. 53 and 56. However, the grain boundaries displayed in these photographs were not generated by any stress applied on the crystal, but formed during the growth process. The mechanism of grain-boundary formation described here is the same as that for the twin formation and for the introduction of the stacking faults found in poly-GeOPc crystals.

5 Structure of Intermediates in Solid-State Reactions

The advantage of high-resolution electron microscopy over X-ray diffraction methods is its possibility to analyze the structure of an aperiodic material or of a defect structure in a crystal. As proved in the previous sections, electron diffraction patterns can provide useful information in the study of reaction mechanisms. This section describes how these techniques can reveal a reaction mechanism by observation of changes in the molecular arrangements caused by the reaction.

5.1 Iodination of Ni–Pc [35]

Doping, which means the incorporation of a different kind of atoms or molecules into a crystal, sometimes improves an electric or other physical property of the crystal [36]. The incorporation of a foreign molecule (or atom) first of all induces a local change in structure, with subsequent changes of the physical and chemical properties of the crystal which are related to this structural change. However, till now no one knew how such changes really look like in a given crystal. High-resolution electron microscopy has revealed the local changes in the structure of Ni–Pc and poly-GeOPc when doped by exposing to iodine vapor. At the final state of the reaction the components form a *stoichiometric* CT-complex and in this point this reaction differs from the so-called "doping" of inorganic semiconductors. However, at its initial state the reaction seems to conform with the general doping mechanism. The reaction of Ni–Pc with iodine has recently been reported by Schramm et al. (1980) [37]. They have analyzed the crystal structure of the stoichiometric complex whose composition is Ni–$PcI_{1.0}$. At the same time they showed that doping with a small amount of iodine also made the crystal of Ni–Pc very highly conductive. The conductivity, however, did not show any remarkable dependence on the iodine content (x). The crystal structure analysis of this stoichiometric complex (x = 1.0) reported by Schramm et al. (1980) yields the unit cell dimensions of this crystal as a = b = 13.936 Å, c = 6.488 Å and the space group as P4/mmc. Ni–Pc molecules are stacked so as to form molecular columns which run parallel to the *c*-axis. The molecular plane of each Ni–Pc is perpendicular to the column axis and differs from that in the original untreated Ni–Pc crystal. An electron micrograph of the original film of Ni–Pc, epitaxially grown on mica, and its high-resolution electron diffraction pattern are shown in Figs. 58a and b. They show that the film is composed of laminar crystallites oriented in three directions; the long axis of each crystal is the *b*-axis whose lattice period is 3.8 Å. The planar molecules of Ni–Pc are stacked parallel, facing one another and forming molecular columns along the *b*-axis. The molecular plane is slightly tilted against the column axis. The tilting angle on the (100) plane of the crystal obviously can be determined from the high-resolution electron micrograph. The projection images of the molecules are strips and they are inclined by an angle of +26° or −26° with respect to the column axis. An electron micrograph of the Ni–Pc film treated with iodine vapor for several hours at room temperature and its HRED pattern are reproduced in Fig. 59a, b. The HRED pattern, Fig. 59b, shows that the reaction is topotactic, that is, the change in crystal structure occurs without an altering of the orientation of the principal

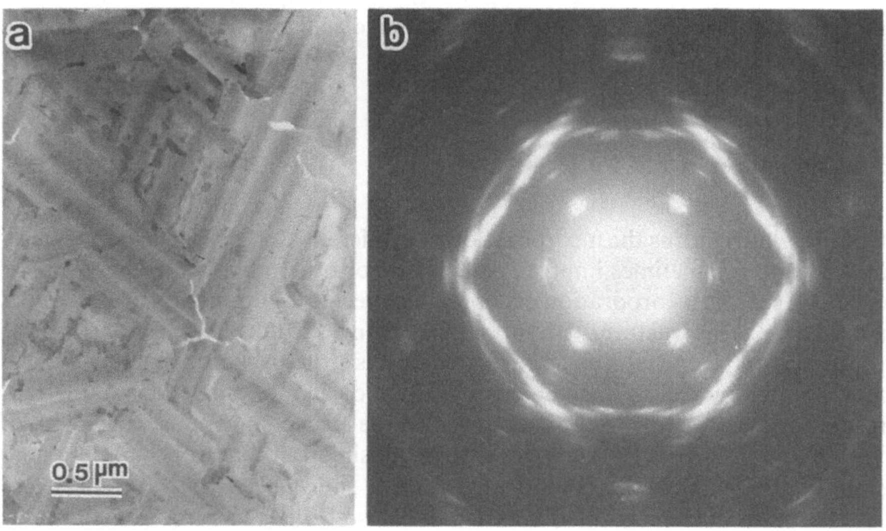

Fig. 58 a, b. Epitaxial film of Ni–Pc grown on muscovite (**a**) and its high-resolution electron diffraction pattern (**b**)

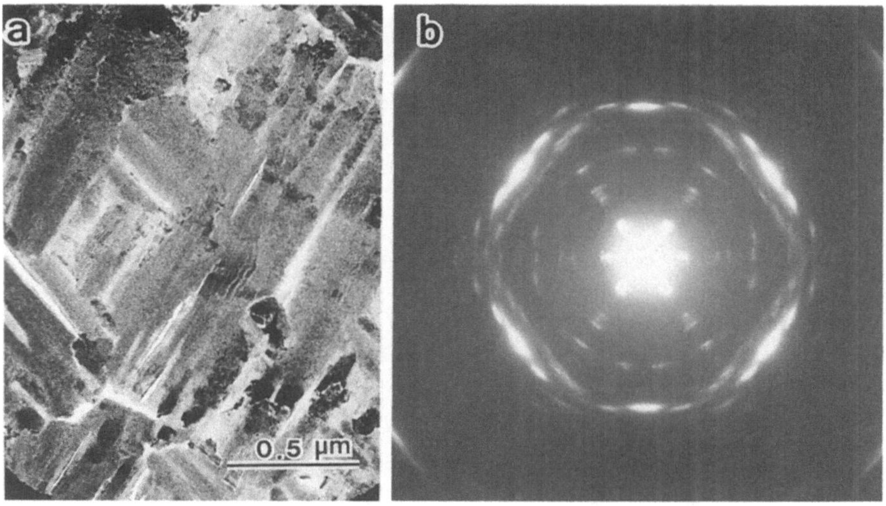

Fig. 59 a, b. Reaction product of Ni–Pc with iodine (**a**) and its high-resolution electron diffraction pattern (**b**). An unreacted part remains and contributes to **b**

axes. The axial correlation between the original and the iodine-treated crystals is illustrated in Fig. 60 where a single orientation has been extracted for simplicity. The fiber period of the crystal has shortened from the original value of 3.8 Å to 3.25 Å by the iodine doping. In compensation, the other lattice dimension perpendicular to the *b*-axis has expanded from 13 Å to 14 Å. In the intermediate stage of the reaction a certain amount of Ni–Pc was left intact. An electron micrograph taken in this intermediate stage is reproduced in Fig. 61. In the area A the lattice interval of 13 Å and the orientation angle of 26° of the molecules remain unchanged, while in the part B lattice fringes of 14 Å interval and a discontinuity and bending of the lattice are indicated. As the tilting angle of the

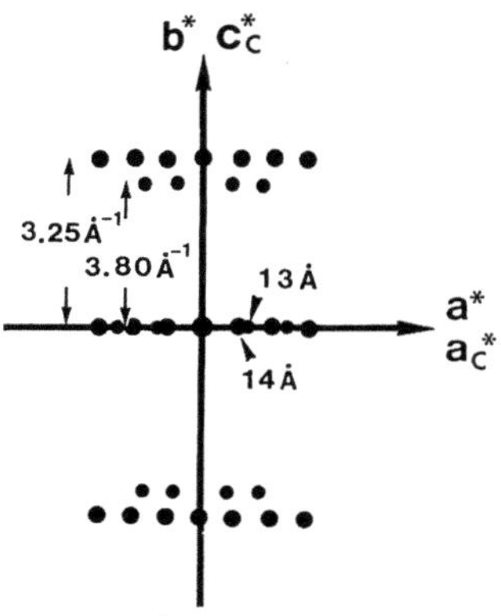

Fig. 60. The axial correlation between the original and iodine doped Ni–Pc in reciprocal space. (The larger dots denote the iodine doped phase, the index c refers to the complex phase)

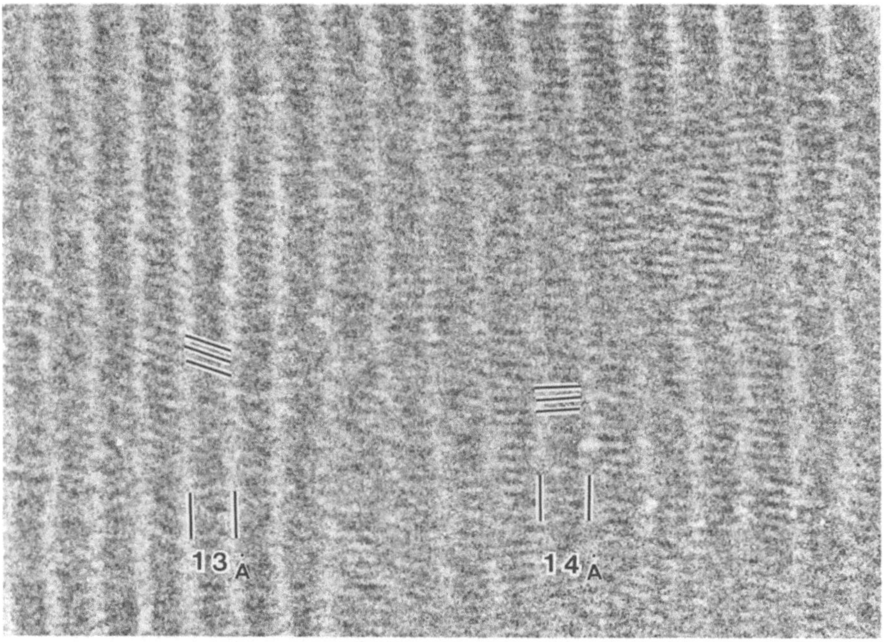

Fig. 61. Molecular arrangement in the reaction intermediate of Ni–Pc and I_2. Molecules are imaged as stripes

molecules in the molecular column decreases on iodine doping, the distance between the centers of neighboring molecules, i.e. the intrastack Ni–Ni distance shortens, while the column-to-column distance expands, as illustrated in Fig. 62. The shortening of the intrastack intermolecular distances causes a discontinuity of the column, and the expansion of the column distances should squeeze the surrounding molecules out of their original positions when the change of the lattice dimensions takes place inside a crystal.

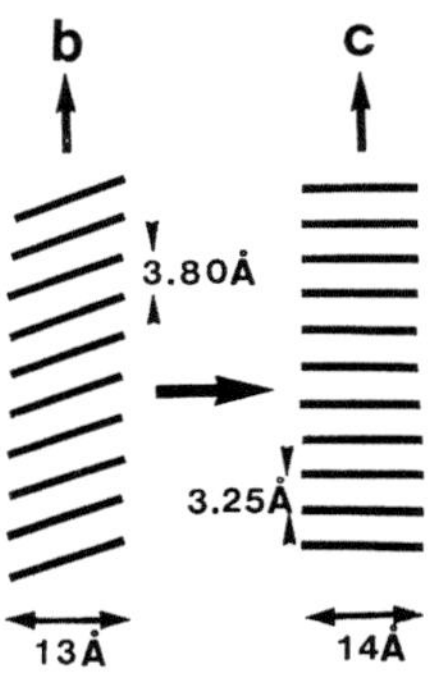

Fig. 62. The change in molecular stacking from the original to the iodine doped Ni–Pc crystals, reacted on muscovite

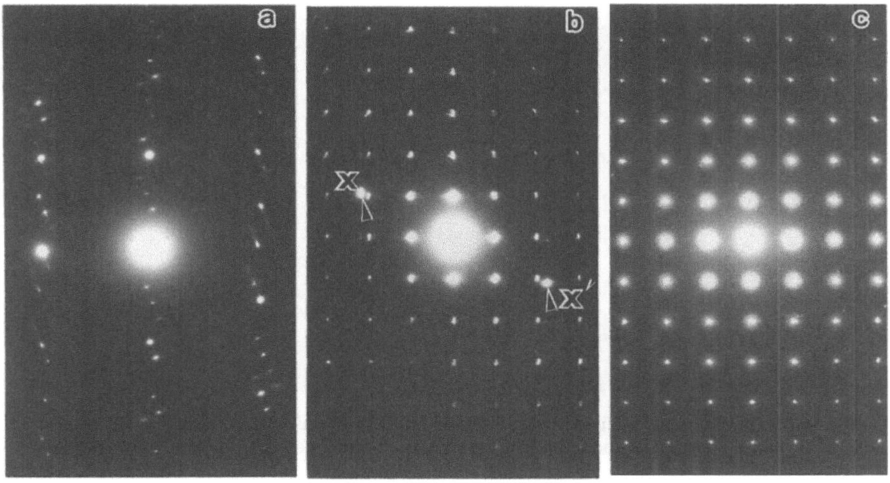

Fig. 63 a–c. Change in the electron diffraction pattern of Ni–Pc due to an uptake of iodine. The initially tilted zone axis becomes normal to the substrate by the iodination, as shown in Fig. 62

Epitaxial Ni–Pc films grown on KCl(001) show the same reaction mechanism as those grown on mica, but the direction of the observation differs from the former case. In the latter film the crystal of Ni–Pc assumes a wedge-shaped dendritic form; its *b*-axis is oblique to the electron beam. The electron diffraction patterns from the Ni–Pc of the original, the intermediate, and the after reaction states with iodine are exhibited in Fig. 63. In this case a topotactic reaction occurs too. However, when the Pc film on KCl is doped, the axis of the molecular column which is oblique in the original film, becomes perpendicular to the substrate KCl (001) face and, therefore, the *c*-axis of the Ni–Pc·I complex formed, that is the molecular column axis of the reaction product, becomes parallel to the incident electron beam, so that the diffraction pattern shows the basal plane of the reciprocal lattice of the newly-formed complex crystal. The periods of the net are $14.0 \times 14.0\,Å^2$, as shown in Fig. 63c. The arrowed spots in (b) correspond to the (402) plane of the unreacted part of the original Ni–Pc film. The axial correlation between the original and the doped crystals is derived from this pattern.

The diffraction pattern indicates that a high-resolution electron micrograph of the crystal can reveal the molecular images as a projection of the columns along the column axis. Figure 64 is the image of such a projection. The [210] direction of the reaction product is almost parallel to the [$40\bar{2}$] direction of the original film of Ni–Pc. This relationship can be represented schematically in real space as shown in Fig. 65. The lines XX′ and YY′ in the figure mark the ($40\bar{2}$) and (210) planes, respectively. The Ni–Pc molecules depicted by dotted lines in Fig. 65b are in contact with the substrate, they keep their original orientation as shown in Fig. 26a over the reaction. Thus, the fact that some molecules keep their original

Fig. 64. Molecular images of a Ni–Pc·I_2 crystal. The Ni–Pc molecules stack normal to the projection plane, but stagger alternately about the projection axis. Iodine was not imaged in this picture

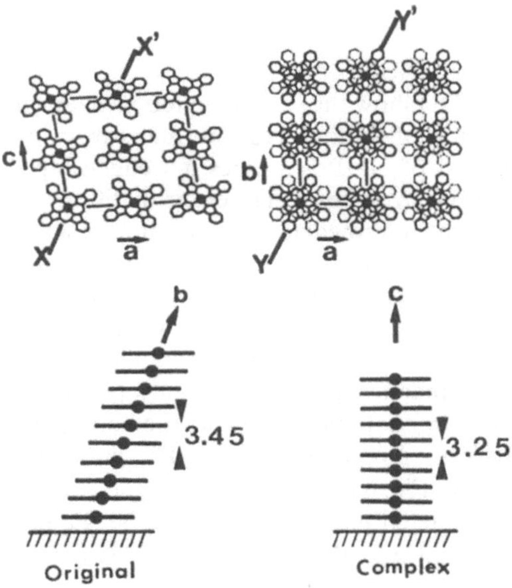

Fig. 65. Structure changes caused by doping iodine into a Ni–Pc crystal film on KCl

orientation and act as a nucleus for the newly produced crystal is an essential feature of a topotactic solid-state reaction. In this case the molecule at the bottom of the molecular column is fixed to the substrate and acts as a nucleus of the new "doped" crystal. As a result, the direction of the molecular columns changes in the reaction. When the substrate KCl is removed, that is, the Ni–Pc is lifted off from the KCl substrate and brought into contact with iodine vapor, the reaction occurs in the same way as was observed with the film grown on mica. The direction of the newly formed molecular columns then does not change from that of the original film.

5.2 Doping of Iodine into Poly-Phthalocyanine [38]

As a representative of low dimensional electrically conductive materials, the family of poly-Pc's has been studied by a variety of physical methods. Most of the poly-Pc's [39] have a similar structure in that metal-Pc's are linked together in a face-to-face orientation by oxygen or halogen atoms lying between the two metal atoms in the center of the adjacent Pc molecules (Fig. 47a). The mechanism of "doping" of these Pc's with iodine, which has been investigated by direct observation of molecular images, occurs as follows: An epitaxial film of GeOPc grown from the vapor phase on KCl (001) was used as a starting material. The film is composed of many crystallites of two well-defined orientations. An electron diffraction pattern of the film shows that the crystallites have the lattice constants of a = b = 13.72 Å, and c = 6.80 Å; they belong to the space group P4/mmc [21]. In the electron diffraction pattern shown in Fig. 66, two sets of the

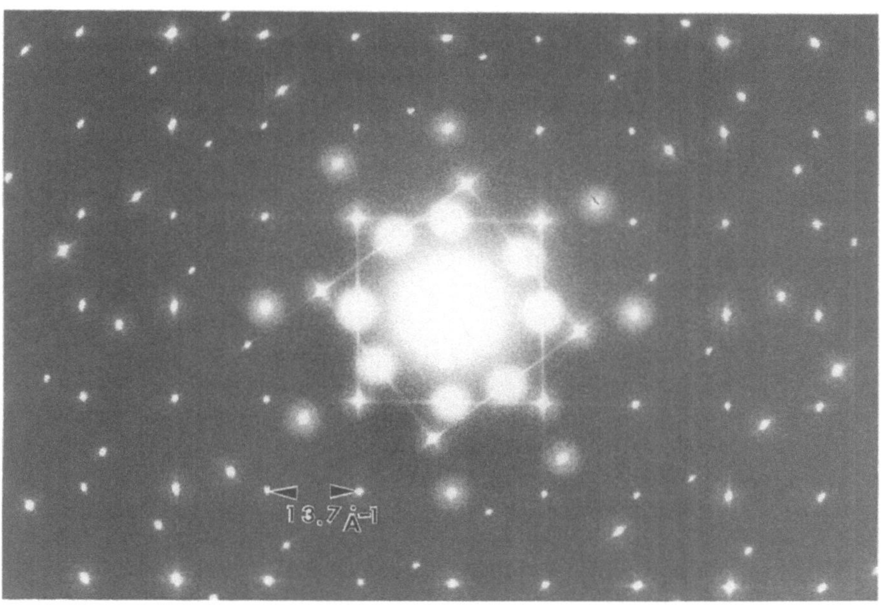

Fig. 66. High-resolution electron diffraction pattern of a poly-GeOPc epitaxial film. Two sets of tetragonal patterns are superimposed with an angle of 37°

reciprocal lattices with these cell dimensions are superimposed on each other with a staggering angle of 37°, i.e. an angle of rotation of a Pc molecule relative to the preceding one in a molecular column. The column is thus composed of molecules stacked with face-to-face orientation with a staggering angle of 37° linked by oxygen atoms which are intercalated between two central Ge atoms. The presence of a square formed by diffuse streaks indicates the existence of stacking faults along the (100) or (010) plane as described above. The molecular columns, identical with the polymer chains, are stacked in a tetragonal array perpendicular to the film plane.

A film formed on a KCl substrate was cut into several small pieces, and placed with a small amount of iodine powder in a Petri-dish. Individual pieces were removed from the dish after different time intervals and the change of the film due to the gradual reaction with iodine vapor was observed by HREM. A HRED pattern of the film at a reaction time of ten minutes is reproduced in Fig. 67. This pattern differs from that of the original film in that diffuse streaks along radial directions occur on the (h00) reflections. The new streaks suggest that the (h00) interplanar distance changes locally on doping. Another difference on iodine doping is the increase of the lattice dimensions. After completion of doping, the unit cell constants of the tetragonal lattice become a = b = 13.90 Å, c = 6.70 Å. That is, the cell size expands along the *a*- and *b*-axes. The axial directions and

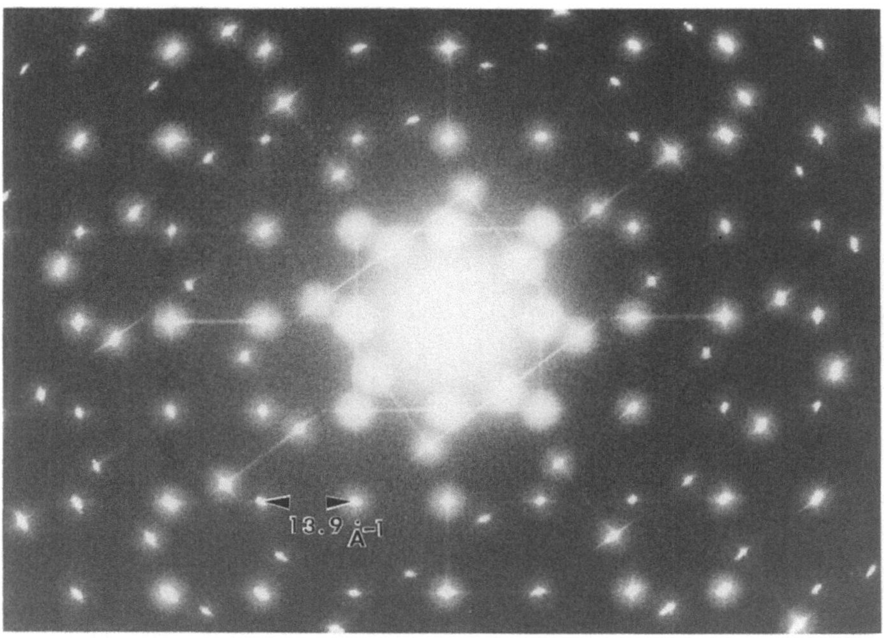

Fig. 67. HRED from poly-GeOPc which was reacted with iodine vapor for 20 min. Streaks appear in radial directions on every (h00) type spot

lattice symmetry are maintained during and after the reaction. High-resolution electron micrographs of the original and doped GeOPc are shown in Fig. 68a, b. Both are photographically processed in order to reduce the photographic noise. Each square in Fig. 68 corresponds to an image of the molecular column of GeOPc projected along the column axis. Every dot at the center of the molecular images (cf. the magnified Fig. 70) is a projection of a row of atoms of —Ge—O—Ge— which bind the Pc molecules together to form the chains. The iodine atoms doped in are located at every corner and surrounded by four Pcs in (b).

The origin of the diffuse streaks running in radial directions on all (h00) and (0k0) diffraction spots is a local change in spacing of these planes, as revealed by the electron micrograph shown in Fig. 69a, b. In the micrograph one can find many cracks along the arrays of square images of the GeOPc molecules. These cracks appear even in a film which was in contact with iodine vapor for only one minute. The width of the crack is constant and is measured to be 16.32 Å including one molecular array, while the spacing without crack is 13.72 Å. Therefore, the expansion width is 2.60 Å. These cleavages cause the streaks of the (h00) reflections perpendicular to the cracks. A part of the image is enlarged in Fig. 69b. It is clear that the cracks always occur along the stacking faults described in a previous section. That is, the molecular arrays facing the crack are displaced along the crack relative to the array on the other side by half the molecular size. This displacement is a stacking fault; many such faults exist, giving rise to the diffuse streaks which form a square in the diffraction pattern. After the

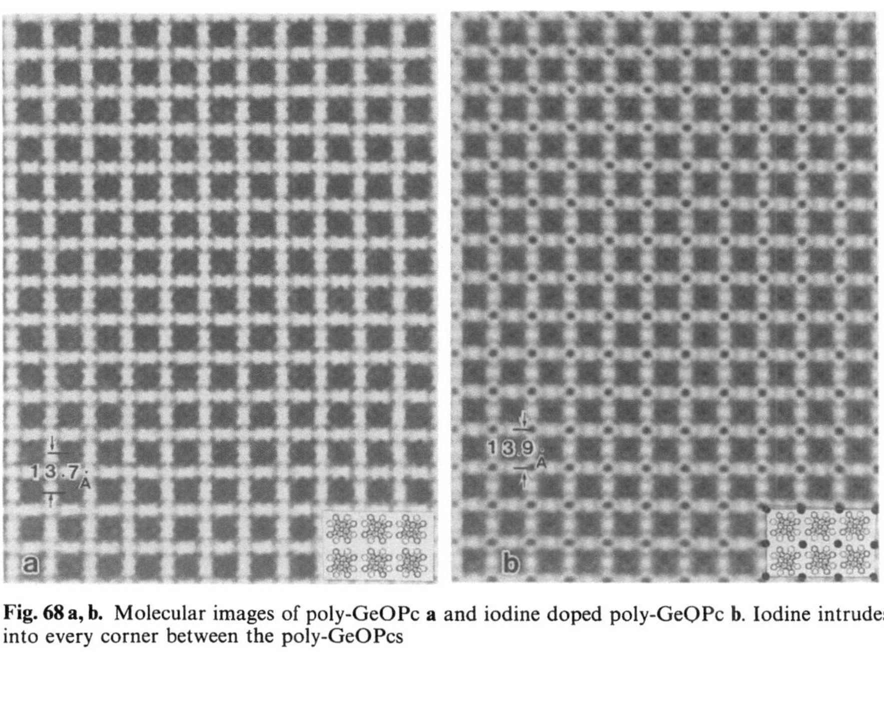

Fig. 68 a, b. Molecular images of poly-GeOPc **a** and iodine doped poly-GeOPc **b**. Iodine intrudes into every corner between the poly-GeOPcs

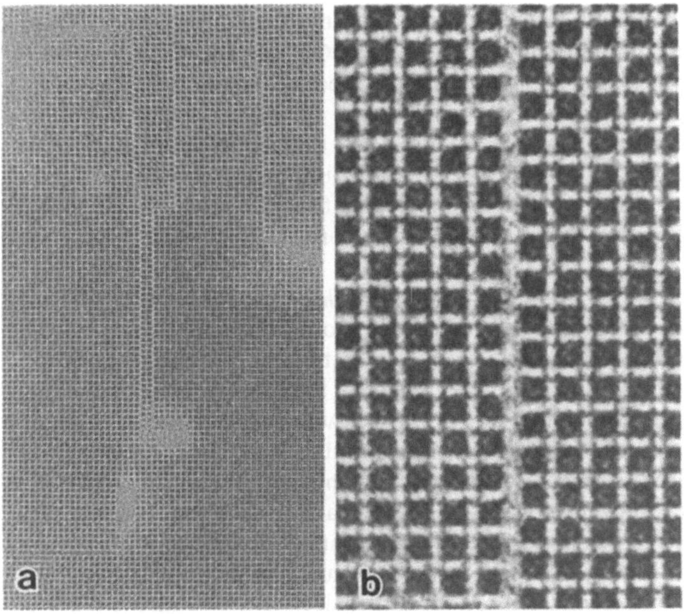

Fig. 69 a, b. Larger area view of a film of poly-GeOPc reacted with iodine vapor **a**, and an enlarged part containing one stacking fault **b**. Crystal cleavage is observed as a white fringe at every faulted plane

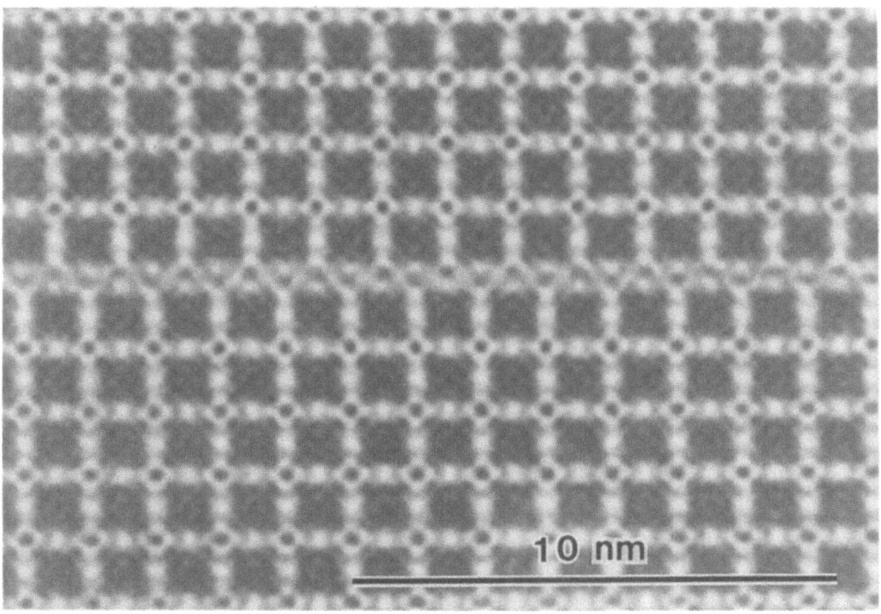

Fig. 70. After the end of the reaction, the iodine position is fixed even at the faulted plane as shown here. The number of iodine atoms is two times more along the fault plane

end of the reaction these cracks are also filled with iodine atoms. This situation is shown in Fig. 70 which includes one stacking fault. The number of the iodine atoms in the fault plane is two times the number at regular sites.

The electron micrograph reproduced in Fig. 71 is that of an intermediate stage of doping and shows that doping goes hand in hand with tearing apart the fault plane. The separation begins from the end of the fault plane (as arrowed) as well as from the middle of such faulted plane (as indicated by an arrow). At this stage of doping, the images of iodine atoms are not definite because the positions of iodines are not yet fixed. The reason why a separation of the interplanar distance as large as 2.6 Å occurs at the fault plane on the uptake of iodine can be explained as follows. Figure 72a represents a space-filling model of the molecules. Van der Waals radii of the carbon and hydrogen atoms are taken from the table of periodic properties of elements published by Sargent as 0.77 Å and 0.37 Å, respectively. The germanium atom is placed at the unit cell origin. At the center of the cell exists a square channel of about 3.5×3.5 Å^2. For the iodine to be taken up by the crystal, the size of the channel has to be somewhat enlarged which leads to the slight increase in lattice dimensions on the reaction. This expansion was observed to amount to 0.2 Å in both directions, i.e. along the *a*- and *b*-axis. Therefore, it can be said that iodine needs a channel of 3.7×3.7 Å^2 in the Pc crystal in order to form the GeOPc-iodine complex.

At the stacking fault the channel becomes smaller as a result of the displacement of the molecular array along the fault plane. This situation is also schematically illustrated in Fig. 72b. Clearly the channel among the molecules becomes smaller and is divided into two parts, one at the center and one at the edge of the unit cell. In order to form a channel of sufficient size for iodine, the plane of the stacking fault has to expand to the same separation as obtained above. When the crystal is cleaved along a stacking fault with a width of 2.6 Å, as measured on an electron micrograph shown in Fig. 67, the channel width reaches this value (3.7 Å), as demonstrated in Fig. 73. Although the required expansion of the lattice is thus larger at the faulted region than in the ordinary regions of the crystal, the penetration of iodine first takes place from the stacking fault.

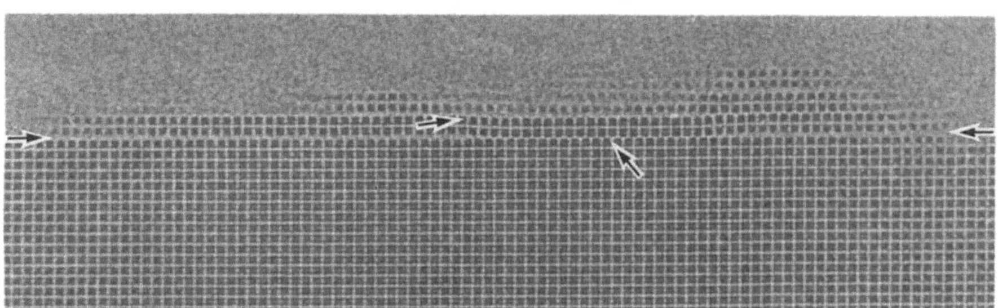

Fig. 71. Electron micrograph of the reaction intermediate of poly-GeOPc·I_2 showing that the cleavages caused during iodine doping are initiated at the stacking faults

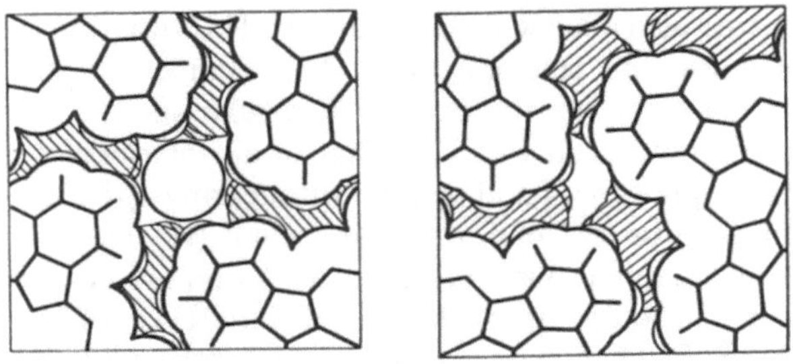

a b

Fig. 72 a, b. The channel existing in the original poly-GeOPc at every corner among the molecules is 3.5 Å × 3.5 Å in size (a) while at the stacking fault plane the channel becomes smaller (b)

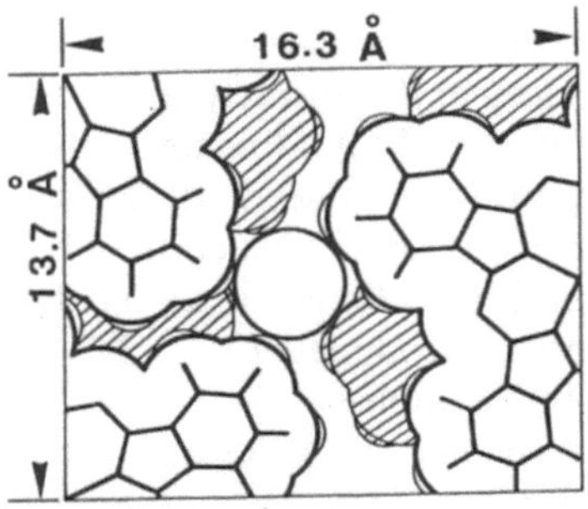

Fig. 73. By expansion of spacings by 2.6 Å iodine can intrude into the voids at the stacking fault as shown here

Presumably cohesion forces acting among the molecular arrays are weaker between the planes facing a stacking fault than in the unperturbed lattice.

The structural lattice defects of organic crystals can be defined by the rotation of the molecules about their center and the lateral shift vector from the regular lattice site. The examples of defects which were visible in the molecular images shown in this article are all unique cases in the sense that the defect structures can be interpreted from the images in a very straightforward way. For most of the organic crystals composed of more complex nonplanar molecules, such an intuitive interpretation is difficult, if not impossible, because the electron micrographs are essentially a two-dimensional projection of the three-dimensional lattice. Reconstructions of three-dimensional images are being tried by the aid of a computer for arbitrary crystals and I believe that such methods will be available in future for the study of defects. By accumulating more data about the defect structures, we can expect fruitful results which will reveal more clearly the role that intramolecular cohesion forces play in the processes of defect formation and crystal growth. Moreover, the study of reaction mechanisms of organic solids in solid-state chemistry and an understanding of the bearing of defects on the physical properties of molecular crystals will greatly benefit from direct lattice imaging by high-resolution, high-energy electron microscopy and electron diffraction in future.

6 References

1. Menter JW (1956) Proc Roy Soc A236: 119
2. Kittel C (1953) Introduction to solid-state physics. Wiley, New York
3. Reimer L (1984) Transmission electron microscopy. Springer, Berlin Heidelberg New York, Tokyo
4. Cowley J (1975) Diffraction physics. North Holland Publ CO, Amsterdam
5. Uyeda N, Kobayashi T, Ishizuka K, Fujiyoshi Y, (1978–79) Chem Scripta 14: 7
6. Born M, Wolf E (1970) Principles of optics, 14th edn, Pergamon Press, New York
7. Glaeser RM (1975) Physical aspects of electron microscopy and microbeam analysis Siegl BM, Beaman DR (eds) Wiley, New York, Chap. 12
8. Williams RC, Fisher HW (1970) J Mol Biol, 52: 121 Fujiyoshi Y, Kobayashi T, Ishizuka K, Uyeda N, Ishida Y, Harada Y (1980) Ultramicroscopy 5: 459
9. Thon F (1971) Electron microscopy in material science. Valdre U (ed) 570, Academic Press, New York
10. Boonstra EG, Herbstein FH (1963) Acta Cryst 16: 252
11. Kistenmacher TJ, Philips TE, Cowan DO (1974) Acta Cryst B30: 763
12. Yase K, Okumura O, Kobayashi T, Uyeda N (1984) Bull Inst Chem Res, Kyoto Univ 62: 242
13. Kobayashi T, Fujiyoshi Y, Uyeda N (1982) Proc 10th Intl Congr Elect Micros Hamburg 1, p 405
14. Uyeda N, Kobayashi T, Ishizuka K, Fujiyoshi Y, Inokuchi H, Saito G (1985) Mol Cryst Liq Cryst 125: 103
15. Kobayashi T, Fujiyoshi Y, Iwatsu F, Uyeda N (1981) Acta Cryst A37: 692
16. Matthews JW (1975) Epitaxial growth. Academic Press, New York
17. Bragg WL (1937) Atomic structure of minerals. Cornell Univ Press, p 210
18. Ashida M, Uyeda N, Suito E (1966) Bull Chem Soc Japan 39: 2616
19. Suito E, Uyeda N, Ashida M, Yamamoto K (1966) Proc Japan Academy 42: 54
20. Ashida M, Hamada A, Watanabe T (1972) Bull Chem Soc Japan 45: 2312
21. Kobayashi T, Uyeda N (1987) J Cryst Growth 84: 5891
22. Uyeda N, Murata Y, Kobayashi T, Suito E (1974) J Cryst Growth 26: 267
23. Vainshtein BK (1964) Structure analysis by electron diffraction. Translated by Feigel E and Spink JA, Pergamon Press, New York
24. Inokuchi H (1960) Bussei 1: 2 (in Japanese)
25. Murata Y, Kobayashi T, Uyeda N, Suito E (1974) J Cryst Growth 26: 187
26. Long RE, Sparks RA, Trublood KN (1965) Acta Cryst 18: 932; Williams RM, Wallwork SC (1968) Acta Cryst B24: 168
27. Uyeda N, Kobayashi T, Ishizuka K, Fujiyoshi Y (1980) Nature 285: 95
28. Amelinckx S (1964) The direct observation of dislocations solid-state physics. Suppl 6, Academic Press, New York
29. Williams J, Thomas JM, Williams JO, Hobbs LW (1974) J Chem Soc Farad Trans II, 71: 138
30. Yase K, Kobayashi T, Uyeda N (1985) Bull Inst Chem Res Kyoto Univ 63: 383
31. Brown CJ (1968) J Chem Soc (A), p 2494
32. Kobayashi T, Fujiyoshi Y, Uyeda N (1982) Acta Cryst A38: 356
33. Kobayashi T, Fujiyoshi Y, Uyeda N (1982) J Cryst Growth 65: 511; Zhang WP, Hou YF (1986) Proc XIth Int Cong Electron Micros Kyoto 1: 829
34. Smith DJ, Fryer JR, Camps RA (1986) Ultramicros 19: 27
35. Kobayashi T, Yase K, Uyeda N (1984) Acta Cryst B40: 26
36. Schramm CJ, Stojakovic DR, Hoffman BM, Marks TJ (1978) Science 200: 47
37. Schramm CJ, Scaringe RP, Stojakovic DR, Hoffman BM, Ibers JA, Marks TJ (1980) J Am Chem Soc 102: 6702
38. Kobayashi T, Uyeda N (1987) Phil Mag B57: 493
39. Inabe T, Gaudiello JG, Moguel MK, Lyding JW, Burton RL, McCarthy WJ, Kannewurf CR, Marks TJ (1986) J Am Chem Soc 108: 7595

Structure, Stability and Crystal Growth of Polymorphs and Polytypes of Long-Chain Aliphatic Compounds

Kiyotaka Sato[1] **and Masamichi Kobayashi**[2]

[1] Faculty of Applied Biological Science, Hiroshima University, Higashi-Hiroshima, 724 Japan
[2] Department of Macromolecular Science, Faculty of Science, Osaka University, Toyonaka, 560 Japan

This article reviews recent studies on crystal structure, thermodynamic stability and crystal growth of polymorphs and polytypes of long-chain aliphatic compounds. Particular attention is paid to the n-fatty acids, since numerous studies have recently been made on these compounds. Details of their molecular structures are described, based on X-ray crystallographic and vibrational spectroscopic data. Using stearic acid, an outstanding model substance, the thermodynamic stability of the different polymorphic as well as polytypic modifications is studied, by precisely measuring the solubility data of each modification (using single crystals). Calculations of molecular interactions combined with vibrational spectroscopic data are also used to evaluate the thermodynamic stability. A few examples of solid-state transformations, where polymorphic and polytypic conversions occur at the same time, are examined by electron microscopic and vibrational spectroscopic methods. The characteristics of nucleation probability and crystal growth rate of different polymorphs and polytypes from various organic solutions are subsequently described. It is found that, depending on the polymorph, the effects of temperature, supersaturation and solvents appear in a different manner. Finally thin film growth by vacuum deposition is studied by electron microscopic observation and analyzed with respect to the importance of the molecular orientation and the underlying vapor growth mechanisms.

1 Introduction

All long-chain aliphatic compounds, like *n*-paraffins, *n*-fatty acids, *n*-alcohols, crystallize in a lamellar structure in which the long-chains are packed almost parallel to each other. The number of crystalline forms which can occur is usually more than one. These forms are classified as belonging to either polymorphism or polytypism [1].

Polymorphism may be defined as the ability to crystallize in different crystal structures whose unit cells differ along the three axes. It may be stressed that numerous long-chain aliphatic compounds reveal rather complicated polymorphism. This is attributed to variable combinations of the different possible molecular interactions which lead to different structures. The representative interactions occur through CH_3 end groups (end-group packing) between adjacent lamellae, and through hydrocarbon planes (side packing) inside individual lamellae. These two principal types of interactions are an essential feature of all long-chain aliphatic compounds. In addition, the interaction through polar groups such as the —COOH group in *n*-fatty acids, or —OH in *n*-alcohols is not negligible.

Polytypism in the long-chain compounds is caused by different stacking sequences of the lamellae along the chain axis, while keeping the two-dimensional arrangements within the individual lamella unchanged. The lamella-lamella stacking is primarily determined by the packing of the end-groups which are usually localized closed to the lamellae interface.

The polymorphism and polytypism of the long-chain aliphatic compounds have been elucidated for reasons of fundamental crystallographic interest. Another point of interest has been that the polymorphism of these compounds has direct bearing on their utilization in industry. Thus, e.g., when the long-chain compounds are used as solid constituents in cosmetics or pharmaceuticals, the physical properties of these final products are related to the dominant polymorphic forms of the constituents. Boistelle [2] recently reviewed data of the crystal structures, occurrence and stability, crystal morphology, lattice defects and crystal growth kinetics of principal *n*-alkanes. Since then, remarkable progress has been attained with elucidating the vibrational spectroscopic properties, thermodynamic stability, occurrence and crystal growth, transformation mechanism and vapor growth of the polymorphs and polytypes of several relevant representatives, most intensively of the *n*-fatty acids. Therefore, the present paper tries to give a brief review on these developments in the past several years.

2 Molecular Structures of Polymorphs and Polytypes

2.1 Crystal Structures

2.1.1 Subcells

The polymorphic structures of the long-chain compounds are characterizable by a subcell which describes the side packing arrangement of the neighboring

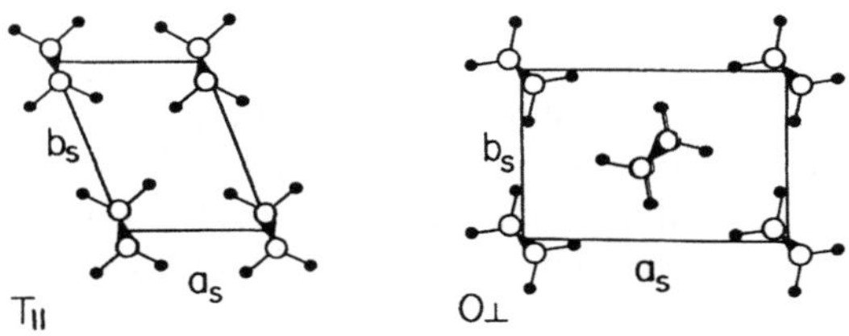

Fig. 1. The Subcell Structures, $T_{\parallel}$ and $O_{\perp}$

molecules in the individual lamella [3]. The subcell is usually constructed by vectors a_s and b_s which describe the two-dimensional lattice perpendicular to the chain axis, and by c_s which equals two hydrocarbons with a distance of approximately 2.54 Å. (Fig. 1). All possible subcells were discussed thoroughly by Abrahamsson et al. [4] and more recently by Hernqvist [5]. The subcells of the crystals consisting of linear long-chains and simple functional groups, such as —COOH or —OH, may be symbolized by $M_{\parallel}$, $T_{\parallel}$, $O_{\perp}$, etc., where M, T and O mean monoclinic, triclinic and orthorhombic crystallographic coordinate systems, respectively. These coordinate systems are followed by the symbols. ∥ and ⊥, which indicate the parallel and perpendicular arrangements of the zig-zag planes of the adjacent hydrocarbon chains.

The triclinic $T_{\parallel}$ subcell in Fig. 1 is known to be one of the most closely packed structures, containing two CH_2 groups in space group symmetry P1; in *n*-octadecane [6] its dimensions are $a_s = 4.528$ Å, $b_s = 5.414$ Å, $c_s = 2.53$ Å, $\alpha = 80.8°$, $\beta = 112.2°$, $\gamma = 121.78°$. The values of a_s, b_s, and c_s do not change appreciably among the different compounds, but the subcell angles vary to a larger extent. This is particularly manifested with the *n*-alkanes, here the three angles change on altering the number n_c of CH_2 units [7]. The most common orthorhombic subcell is $O_{\perp}$, which contains the hydrocarbon chains with two symmetry-related orientations of the chain planes (Fig. 1). This subcell with four CH_2 groups belongs to the space group Pnam, with dimensions $a_s = 7.40$ Å, $b_s = 4.93$ Å and $c_s = 2.53$ Å, (with a choice of the axes as utilized in spectroscopic studies in polymer science). Another cell choice of $O_{\perp}$, Pbnm, in which the vectors of a_s and b_s are interchanged with respect to the present choice, has been employed mainly in crystallographic studies [4, 5].

Besides these two subcells, less commonly observed subcells are $M_{\parallel}$, $O_{\parallel}$, $O'_{\parallel}$ and hybrid subcells occurring in crystals of rather complicated molecules, as e.g. in those having more bulky functional groups ($M_{\parallel}$, $O_{\parallel}$, $O'_{\parallel}$ and hybrid subcells) or cis-double bonds ($O'_{\parallel}$) [4, 5]. Furthermore, a hexagonal subcell (H) describes the molecular packing of the disordered state (rotator crystalline phase) of some *n*-alkanes, which appears just below the melting point.

Some typical crystal structures revealing the $T_{\parallel}$ and $O_{\perp}$ subcells are summarized in Table 1. The stability and occurrence of different crystalline structures for a given series of compounds may be of great interest. Key parameters of solving this problem may be the number n_c and the parity (even or odd number of carbon atoms) of the hydrocarbon as well as the crystallization conditons.

Table 1. The subcell and unit cell structures of some principal long-chain compounds

Subcell	Compound	Formula	Unit cell (space group)	Ref.
	n-Octadecane	$C_{18}H_{38}$	triclinic ($P\bar{1}$)	[6]
	Lauric acid A-super form	$C_{12}H_{26}O_2$	triclinic ($A\bar{1}$)	[8]
$T_{\parallel}$	Lauric acid A_1 form	$C_{12}H_{26}O_2$	triclinic ($P\bar{1}$)	[9]
	Tridecanoic acid: A′ form	$C_{13}H_{28}O_2$	triclinic ($P\bar{1}$)	[10]
	Hexatriacontane: orth. I form	$C_{36}H_{74}$	orthorhombic ($Pca2_1$)	[11]
	Hexatriacontane mon. form	$C_{36}H_{74}$	monoclinic ($P2_1/a$)	[12]
$O_{\perp}$	Tritriacontane orth. form	$C_{33}H_{68}$	orthorhombic (Pcam)	[13]
	Stearic acid: B form	$C_{18}H_{38}O_2$	monoclinic ($P2_1/a$)	[14]
	Stearic acid: C form	$C_{18}H_{38}O_2$	monoclinic ($P2_1/a$)	[15]
	Heptadecanoic acid: B′ form	$C_{17}H_{36}O_2$	triclinic ($P\bar{1}$)	[16]
	n-Hexadecanol	$C_{18}H_{38}O$	Monoclinic ($A2_1/a$)	[17]

2.1.2 Polymorphic Structures

The thermodynamic stability of different polymorphs is determined by the Gibbs free energy of their crystal structure. The formation and existence of the different polymorphs, however, are also dependent on temperature, purity of the sample, method of crystallization, and so on. An approach based on lattice energy calculations may serve to interpret the occurrence and stability of the polymorphs from a crystallographic point of view [2].

In the case of *n-alkanes*, the triclinic structure is favored for $n_c < 12$, irrespective of the parity of the considered hydrocarbon. For longer chain lengths, the odd-numbered *n*-alkanes favor the orthorhombic form, whereas the triclinic or monoclinic forms are found in the even *n*-alkanes. A theoretical approach [2] to partly explain these tendencies can briefly be described as follows. With decreasing n_c, the relative importance of the end-group packing increases. This favors the triclinic structure. For longer chains, the crystal structure may primarily be influenced by the symmetry of the molecule. As the molecule of an odd-numbered *n*-alkane is unsymmetric, the two CH_3 groups protrude in opposite directions. Hence, a vertical structure with respect to the lamella plane is preferred, favoring the orthorhombic symmetry. On the contrary, the two CH_3 groups of the symmetric even-numbered molecule protrude in the same direction, favoring an oblique or a vertical structure. As to the relative stability of the monoclinic and the orthorhombic structure, the calculated lattice energy was found lower in the monoclinic than in the orthorhombic form. However, the difference was rather small so that influences of homologues (as impurities) or different growth conditions may alter the occurrence probability.

Table 2. Polymorphism in *n*-fatty acids

n_c*	Name	Polymorph**	n_c	Name	Polymorph
18	Stearic acid	A(t), B(m), C(m), E(m)	17	Heptadecanoic acid	A′(t), B′(t), C′(m)
16	Palmitic acid	A_2(t), B(t), C(m), E(m)	15	Pentadecanoic acid	A′(t), B′(t), C′(m)
14	Myristic acid	A_2(t), B(t), C(m)	13	Tridecanoic acid	A′(t), C′(m)
12	Lauric acid	A_1(t), C(m) A-super (t)			

* n_c: number of carbons of aliphatic chains.
** t: triclinic, m: monoclinic.

The polymorphism of the *n-fatty acids*, however, is quite complicated in comparison to *n*-alkanes, as may be seen from some principal saturated fatty acids which have been analyzed by modern crystallographic methods (Table 2). A general tendency is that more polymorphs occur among the even-numbered acids than among the odd-numbered acids. For example, palmitic acid possesses four polymorphs. In contrast, *n*-heptadecanoic acid with $n_c = 17$, possesses three forms. In addition, the different polymorphs crystallize simultaneously from solvent, (this is particularly pronounced for the even-numbered acids), and do not convert to other forms at ambient temperature. These characteristics are ascribed to specific features of the crystallization and polymorphic transformation processes of the *n*-fatty acids.

The multiplicity in the polymorphism of the *n*-fatty acids may be discussed in terms of the molecular interactions involving the endgroup packing, the aliphatic side packing and the carboxylic interactions. The former two factors may play the same role as in the *n*-alkanes described above. A judgement of the relative importance of the carboxylic interaction, however, is critical since there are two molecular configurations, *cis* and *trans*, caused by a proton transfer in the cyclic carboxylic group of the acid dimer. The relative stability of the two configurations is found to be strongly dependent on the parity of n_c and on the respective polymorphic structure (see Sect. 2.2). Hence, subtle differences in the molecular structures between the polymorphs often appear either in the aliphatic chain portions close to the —COOH group or in the arrangement of the —COOH group itself. No general model, however, has been successful to fully interpret the stability of the complicated polymorphs of the *n*-fatty acids. Instead, their stability has been determined experimentally by measuring the solubility of each modification at different temperatures, using stearic acid as a model substance (see Sect. 3).

Complete structural analyses of the polymorphs of the *n*-alkanes are described in a previous review [4]. All crystals contain the aliphatic chains in all *trans* conformation except in the rotator phase. The representative crystal structure is either orthorhombic (e.g., *n*-$C_{23}H_{48}$) [18], or monoclinic (*n*-$C_{36}H_{74}$) [12], or triclinic (*n*-$C_{18}H_{38}$) [6]. The variations in the crystal structures of the *n*-alkanes can be classified into these three, since the polymorphism in the *n*-alkanes

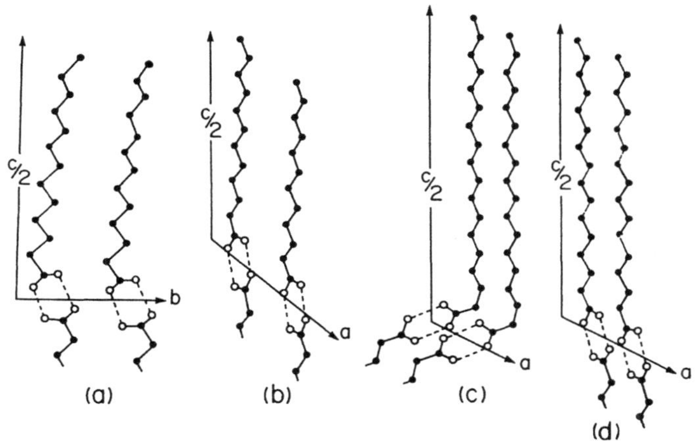

Fig. 2 a–d. Molecular structures of the unit cells of (**a**) the A′ form of tridecanoic acid, (**b**) the C form of lauric acid, (**c**) the B form of stearic acid and (**d**) the E form of stearic acid

is rather simple. In contrast, the polymorphic structures of the *n*-fatty acids differ more widely.

Figure 2 shows the crystal structures of four representative polymorphs of the *n*-fatty acids crystallizing with $T_{\parallel}$ and $O_{\perp}$ subcells. The lattice parameters are summarized in Table 3; to discuss the respective crystal structure, the relation between the axes of the subcell and those of the unit cell is also given.

As to $T_{\parallel}$, three forms, A′, A_1 and A-super, were precisely analyzed. In addition, a form A_2 was isolated in palmitic and myristic acid with spectroscopic methods [19]. In the A′ form of tridecanoic acid (Fig. 2a) [10], one unit cell contains two molecules. The molecular conformation in the aliphatic chain is all-*trans*. The carboxyl and methyl groups are situated in alternative layers, forming a simple layer structure. The same structure was observed in the A′ form of pentadecanoic acid [20]. The layer structure of A_2 is identical to A′, but its unit cell contains two types of dimer chains that are different from each other in the conformation of the carboxyl group [19]. The A-super form of lauric acid assumes a six-fold superstructure [8, 21], containing two types of dimer chains

Table 3. The lattice parameters of typical polymorphs and polytypes of *n*-fatty acids

Form	n_c	a	b (Å)	c	α	β (degrees)	γ	Subcell orientation
A′	13	4.273	4.972	37.66	90.36	109.44	112.43	$a_s \parallel a$
C	18	9.36	4.95	50.7	90	128.15	90	$b_s \parallel b$
B_m	18	5.591	7.408	49.38	90	117.24	90	$a_s \parallel b$
B_o	18	5.58	7.34	87.58	90	90	90	$a_s \parallel b$*

* In Ref. [32], the a and b axes are reversed, hence the space group of B_o was designated as Pbca.

differing in the conformation of the carboxyl group with respect to the aliphatic chain. The A_1 form of lauric acid is formed by dimers, in which the carboxyl group of one dimer is located adjacent to the methyl group of the neighboring dimer [9]. As to the A form of stearic acid, no trial of a precise X-ray analysis has been successful, although it frequently crystallizes from solvent [22].

In the polymorphs having the $O_\perp$ subcell, the C form of stearic acid [15] and lauric acid [23] displays all-*trans* conformation; the unit cell contains two dimers (Fig. 2b). This structure is more or less similar to the linear chain structures of the *n*-alkane crystals. The B form of stearic acid differs remarkably from the C form (Fig. 2c). The molecular conformation is all-*trans* except for the C_2-C_3 segments which reveal *gauche* conformation. This conformation was recently found [14], but not in earlier studies [24, 25]. Another difference between B and C is a tilt of the long chain axis with respect to the lamellar plane; C is more tilted than B. Additionally, the geometric relations between the axes of the subcell and those of the unit cell are also different (Table 3). In B b is parallel to a_s, whereas in C it is parallel to b_s. Accordingly, the long-chain is tilted over the short axis of the $O_\perp$ subcell in B, whereas it is tilted over the long axis in C. This relation is in contradiction to a previous study [15]. The E form, reported for stearic [26] and palmitic acids [27] which coexists with B, is obtained very seldom. The crystal habits of B and E are identical, indicating the same arrangement of the aliphatic packing. However, infrared [26] and Raman [28] spectra indicate that E consists of planar zigzag molecules like C; hence no *gauche* conformation is revealed (Fig. 2d). The space group of E is $P2_1/a$.

Among the $O_\perp$ subcell crystals of the odd-numbered fatty acids, the C′ form of undecanoic acid [29] has the same structure as stearic acid C. However, the B′ form of heptadecanoic acid seems to exhibit a triclinic unit cell [16].

2.1.3 Polytypic Structures

Crystal structures of polytypes of the long chain compounds have so far been reported for n-$C_{28}H_{58}$, n-$C_{36}H_{74}$ [30, 31] and for stearic acid B [32], all of which are polytypic modifications of the monoclinic unit cell with the $O_\perp$ subcell. The crystal form of these polytypes is orthorhombic. If the orientations of the monoclinic layers are symbolically denoted by A and ∀, the resultant polytypes may be described, in general, as a periodic sequence of $(A_p\,\forall_q)$. The crystal forms of the two above mentioned polytypes are the extreme case where $p = q = 1$, in which two adjacent layers in the orthorhombic unit cell are related by a two-fold rotation axis perpendicular to the (001) plane. More complicated polytypic modifications can occur, where the values of p and q are different from 1, or are random numbers in the sequence of layers. Observations of complicated spiral steps on the *n*-alkane crystals indicate the existence of these complex polytypes [33]. However, their detailed crystal structures have not been reported.

Two nomenclatures have been used to describe the orthorhombic polytype: 2O[180˙] [30], according to Ramsdell's designation and *orth II* [32] in

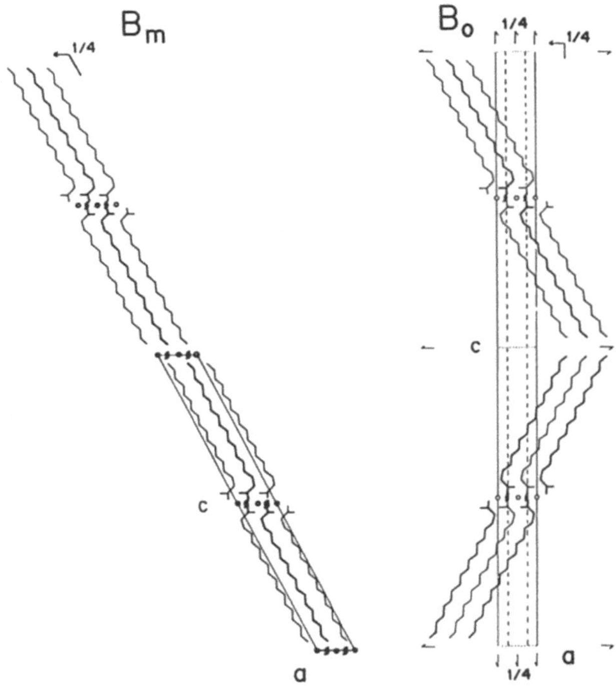

Fig. 3. Unit cell of the monoclinic crystal structures B_m and the orthorhombic polytype B_o of stearic acid

comparison to the normal orthorhombic polymorph (designated as *orth I*) for n-$C_{36}H_{74}$ [11]. The present article uses an abbreviated symbol, index o, together with index m for the normal monoclinic form.

The crystal structure of B_o of stearic acid is shown in Fig. 3[32]. The a_o and b_o dimensions are essentially the same as those in B_m in Fig. 2c, whereas the c_o dimension equals $2c_m \sin \beta_m$. The c_o axis is normal to the lamellar interface. From the observed systematic absence of the X-ray Weissenberg reflections [$k \neq 2n$ for (0kl), $1 \neq 2n$ for (h0l), $h \neq 2n$ for (hk0)], the most possible space group of B_o is Pcab. This structure is essentially the same as that of the orthorhombic polytypes of n-$C_{28}H_{58}$ and n-$C_{36}H_{74}$ [30]. It is noteworthy that particular differences between the orthorhombic and monoclinic polytypes should appear in the methyl group packing at the lamellar interfaces and, more macroscopically, in the vibrational symmetry. Both aspects should be reflected in the vibrational spectra, as will be described in the next section.

In case of the C polymorph, no other form than the normal monoclinic one has been discovered so far.

2.2 Vibrational Spectroscopic Characteristics

Infrared and Raman spectroscopy have been established as powerful tools for investigating various molecular structures in a wide variety of substances. In the case of the long-chain aliphatic compounds, however, comprehensive spectroscopic studies were essentially limited to the *n*-alkanes. This is due to the difficulty of preparing well-grown single crystals, which can be used for polarization measurements which are essential for a detailed analysis of the spectra. A significant progress has been made in this field since several years, owing to successes in preparing single crystals from purified materials, and to developments in various microprobe techniques, used in combination with high-sensitivity spectrometers.

2.2.1 Molecular Conformation and Subcell

The saturated hydrocarbon chains take the all-*trans* conformation in crystalline phases. Molecular vibrations of the infinitely-long all-*trans* polymethylene chain are characterized by the phase angle $\phi(-\pi \leq \phi \leq \pi)$ between the vibrating neighboring structural units. Here, "structural unit" means the translational repeat unit along the chain axis, which is the $—CH_2—CH_2—$ unit for polymethylene. The normal frequency ν for a particular mode varies with ϕ. The function expressing the ν versus ϕ relationship is called a dispersion curve or a frequency branch. The number of the branches is 3N when the structural unit contains N atoms. For the polymethylene chain with $N = 6$, there are 18 branches. Conventionally, these are divided into 9 branches from ν_1 to ν_9 [35–38] with phase angles $\phi' = \phi/2$. The vibration modes at $\phi = 0$ (modulo 2π), or $\phi' = 0$ and $\pm\pi$, are named zone center modes; these are infrared and/or Raman active. Since the frequencies of the lowest 4 branches of a single polymer chain approach zero when $\phi \to 0$ (the acoustic branches), there are 3N–4 zone center modes. Among the 14 zone center modes of the polymethylene chain, the following Raman-active modes are very sensitive to the conformational order in the chain: the antisymmetric CH_2 stretch mode $\nu_a(CH_2)$ ($\sim 2285\,cm^{-1}$), the CH_2 twisting mode $t(CH_2)$ ($\sim 1295\,cm^{-1}$), the CH_2 rocking mode $r(CH_2)$ ($\sim 1168\,cm^{-1}$), the symmetric CC stretch mode ν_s (CC) ($\sim 1130\,cm^{-1}$), and the antisymmetric CC stretch mode ν_s(CC) ($\sim 1060\,cm^{-1}$). In Fig. 4 the Raman spectrum of a highly crystalline polyethylene (PE) crystallized from the melt at 10 kbar (the PE chains are in the fully extended state) is compared with that of molten PE.

In the all-*trans* polymethylene chain of a certain *finite* length, only those molecular vibrations are allowed, which belong to a particular set of ϕ values which depends on the chain length, the type of vibration, and the nature of the chain end. The modes with the lowest ϕ value correspond to the zone center modes of the infinite chain, they are called "the most-in-phase-modes". These modes appear at positions and with intensities very close to those of the

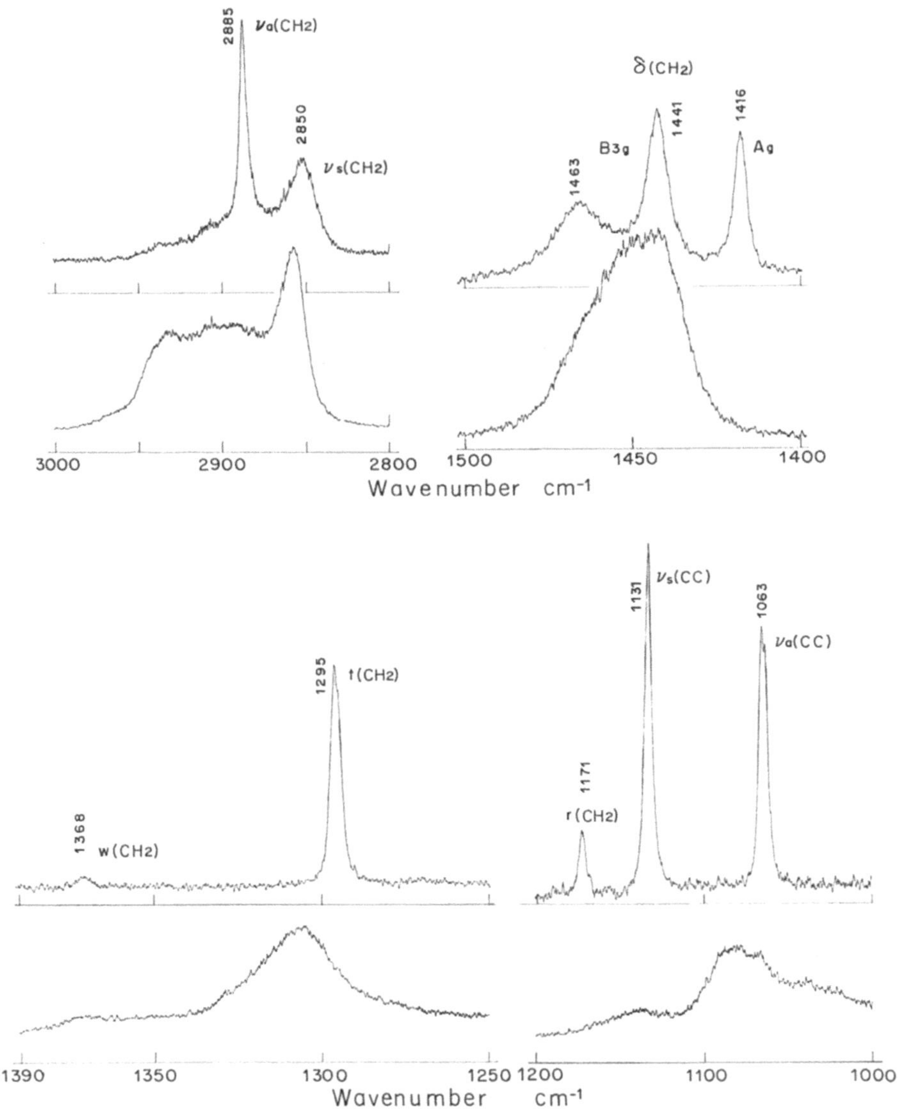

Fig. 4. Raman spectra of highly crystalline (*upper*) and molten (*lower*) polyethylene samples

zone center modes of the infinite chain. For example, the above-mentioned Raman bands are used as a measure of conformational ordering in the hydrocarbon chains. Figure 5 shows polarized Raman spectra of a single crystal of orthorhombic n-$C_{35}H_{72}$ measured at room temperature.

In the polarized Raman spectra, the scattering geometries with respect to the crystallographic axes are expressed in the following manner (Porto's notation;

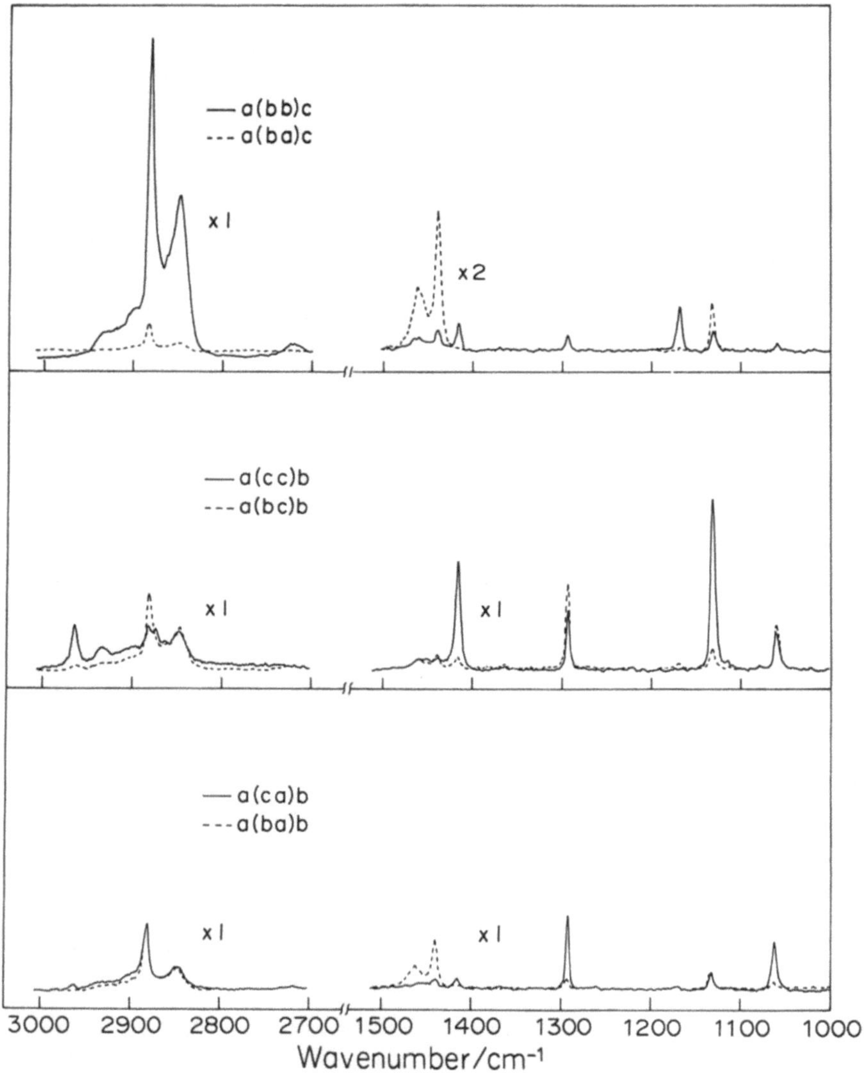

Fig. 5. Polarized Raman spectra of the orthorhombic n-$C_{35}H_{72}$ single crystal taken at room temperature

using a(ba)c as an example; symbols outside the parentheses concern the light propagating directions, (here: incident beam along the a-axis and the scattered light observed in the direction along the c-axis); symbols inside the brackets indicate the polarization directions, (for the incident beam along the b-axis and for the scattered light along the a-axis). The combination of the incident and scattered polarization, say (ba), corresponds to the Raman tensor element α'_{ba} ($= \alpha_{ab}$) which can be measured in the scattering geometries a(ba)c or a(ba)b.

From the data in Fig. 5 we are able to determine the symmetry of the vibrational mode to which the Raman band belongs. For example, the 1410 cm^{-1} band appears in the a(bb)c and a(cc)b geometry, meaning that this band contains the tensor elements of α'_{bb} and α'_{cc}. So, this band is assigned to an A_g mode (a totally symmetric mode). By this way, the vibrational assignments of all of the Raman bands of *n*-alkanes and polyethylene are performed. For the finite chains, in addition to the most-in-phase-modes, some vibrational modes with allowed ϕ values obtain higher intensities, and give rise to band series which are called "progression bands" in particular frequency ranges. The pattern of this band series is very sensitive to the conformation of the hydrocarbon chain [39, 40].

The conformations of other groups are also reflected in the spectra. For example, the conformation around the $—CH_2—CH{=}CH—CH_2—$grouping in unsaturated fatty acids can be detected from the positions of the bands which are caused by the CH stretching mode, the C=C stretching mode, the in-plane and out-of-plane CH bending modes, and the C=C—C bending mode [41, 42]. Finally the *cis* and *trans* configurations of the dimerized carboxyl groups give rise

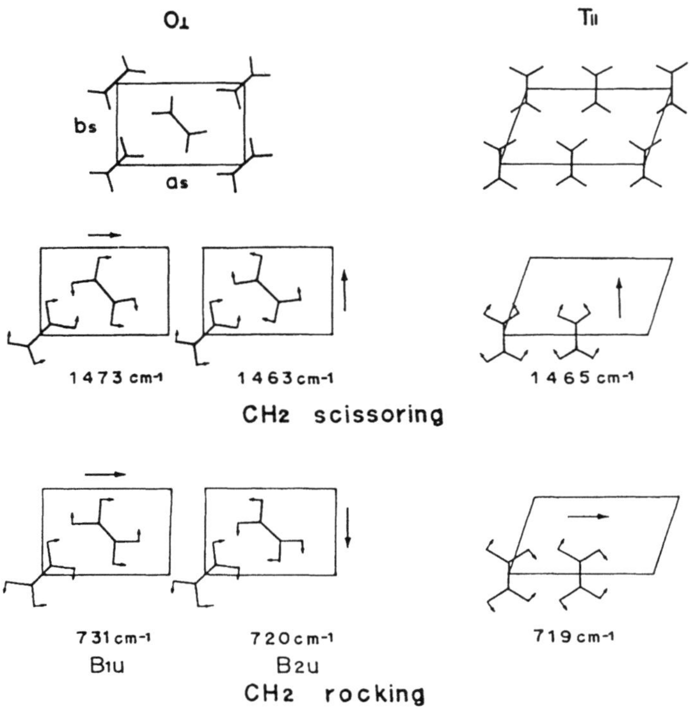

Fig. 6. Methylene scissoring and rocking vibrational modes with their infrared polarization (*bold arrows*) in the $O_\perp$ and $T_\parallel$ subcells

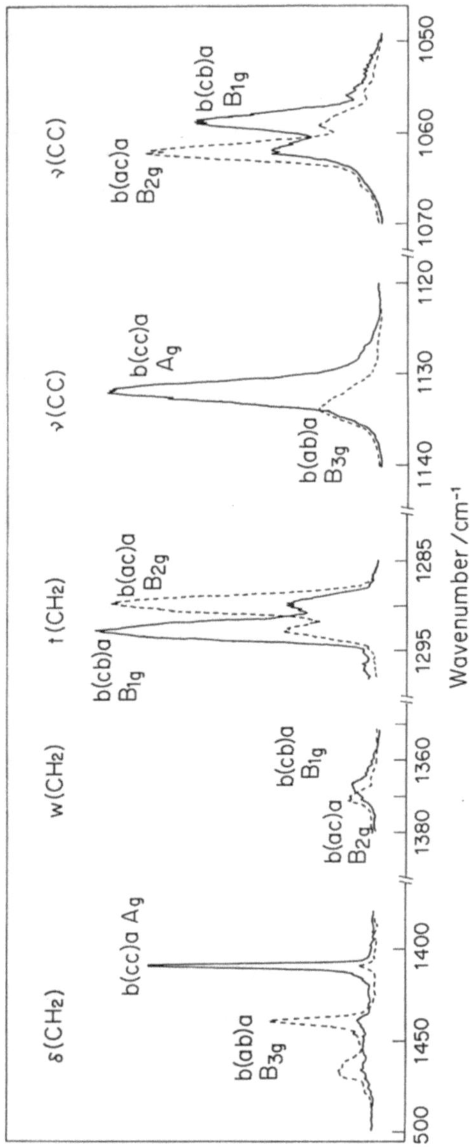

Fig. 7. Splitting of the Raman bands of the orthorhombic n-$C_{35}H_{72}$ single crystal measured at liquid helium temperature

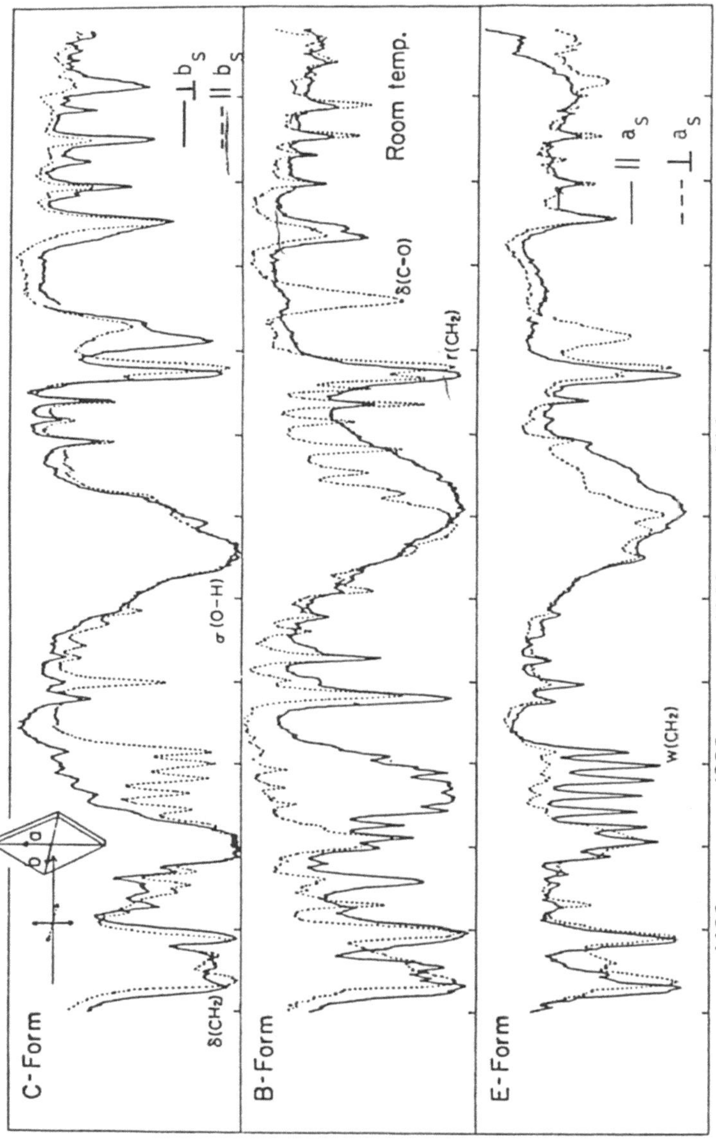

Fig. 8. Polarized infrared spectra of the C, B and E forms of stearic acid at room temperature

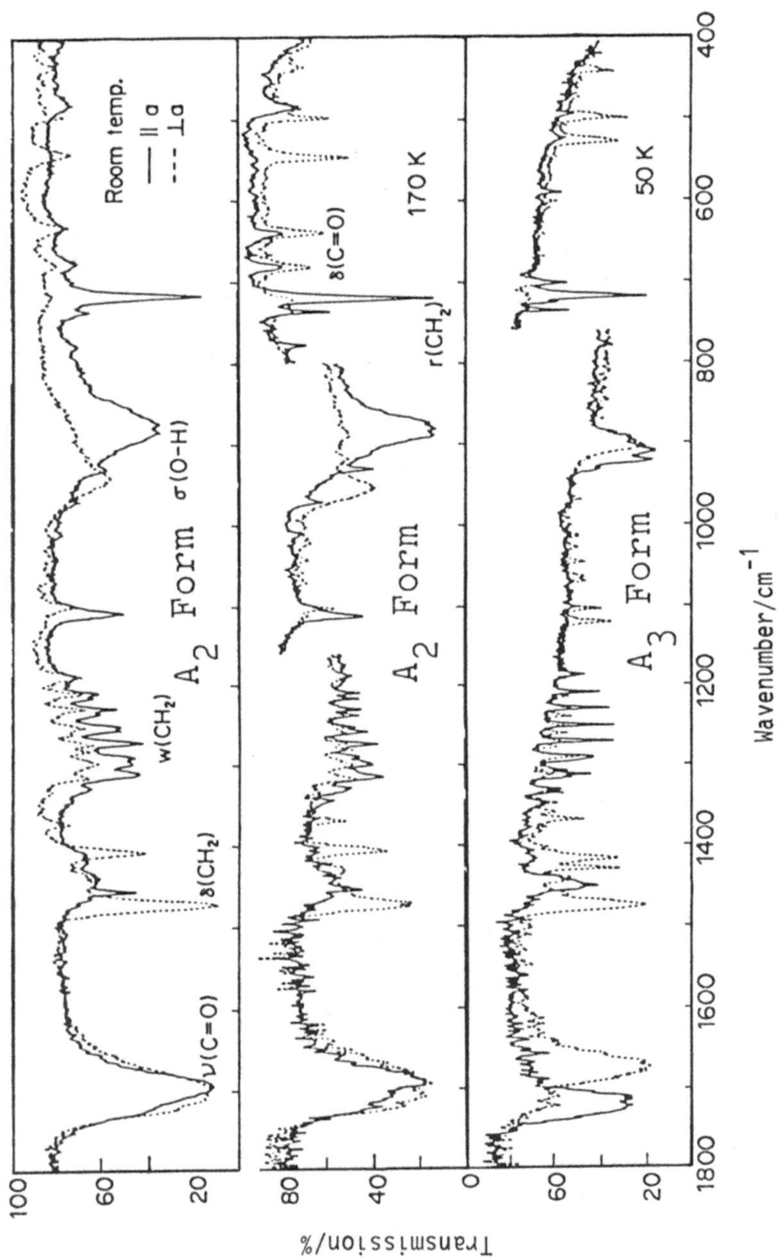

Fig. 9. Polarized infrared spectra of the A_2 and A_3 forms of palmitic acid at room temperature

to bands which are caused by the out-of-plane O—H bending mode δ(OH), and the in-plane C=O bending mode δ(C=O), at characteristic spectral positions, as will be discussed below [43, 44].

As to the spectral characteristics of the subcell structures, the parallel and the perpendicular arrangements provide remarkably different patterns of the band splitting which is caused by different chain-chain interactions. In the $O_\perp$ subcell, the band associated with molecular mode splits into a doublet, which exhibits characteristic dichroism. For example, the r(CH_2) infrared band splits into two components at 731 cm^{-1} (a_s-polarized) and 720 cm^{-1} (b_s-polarized) as shown in Fig. 6. Each band in the ν_8 progression, the r(CH_2)-t(CH_2) branch, also splits into the a_s and b_s polarized components. The ϕ dependence of the split pattern is dependent on the structure of the unit cell [45]. In the Raman spectrum, the $O_\perp$ subcell gives rise to a sharp band at 1416 cm^{-1} associated with a split component of the Raman-active δ(CH_2) mode, which belongs to the A_g species of the space group of the subcell [46]. Splitting of other Raman bands is very small, being hardly detectable at room temperature: the splitting is seen in the ν_a(CC) band of highly crystalline PE, see Fig. 4; at liquid helium temperature it becomes clearly detectable (Fig. 7) [47].

As typical data of the $O_\perp$ subcell, polarized infrared spectra taken with single crystals of the C, B and E forms of stearic acid are shown in Fig. 8 [48]. The δ(CH_2) and r(CH_2) bands split into the a_s- and b_s-polarized components. Figure 9 displays the polarized infrared spectra of the A_2 and A_3 forms of palmitic acid [19], as an example of the $T_\parallel$ subcell. Here, the δ(CH_2) and r(CH_2) bands appear as singlets, polarized perpendicular and parallel to the a_s axis of the $T_\parallel$ subcell, respectively.

For the hexagonal subcell, the spectral pattern is very close to that of $T_\parallel$, but the polarization is completely scrambled within the a_s–b_s plane [49].

2.2.2 Spectral Data of Polymorphs and Polytypes

Both the C and E forms of even-numbered fatty acids consist of all-*trans* hydrocarbon chains packed in the $O_\perp$ subcell. This kind of structural similarity is reflected in their spectral patterns. However, there is a distinctive difference between the two forms in the polarization of the ν_3 progressive band series and also in the bands which are due to the σ(OH) and the δ(C=O) modes of the carboxyl group [26, 27, 48]. This is caused by the difference in the directions of the chain inclination and in the packing of the carboxyl groups. The spectral difference between the A_2 and A_3 forms of palmitic acid can also be ascribed to the differences in the carboxyl packing scheme [19].

Among all polymorphs of even-numbered fatty acids, the C form is thermodynamically most stable in the highest temperature range (see next section). This fact is ascribed to a molecular motion occurring in the carboxyl groups. The carboxyl dimer in the C form is in a dynamic equilibrium between two stable configurations, *cis* and *trans*, which transform by a rotation about the

C_1—C_2 bond. The concentration ratio of the two configurations varies with temperature. The *cis* form is predominant at room temperature, but the *trans* concentration increases with decreasing temperature, approaching 100% at 4 K [43, 44]. The structural change is reflected in the temperature dependence of the intensity ratio of pairs of spectral bands of which one is associated with one, the other with the other of the two configurations. For example, the $\delta(C{=}O)$ bands at 690 and 670 cm^{-1} are due to the *cis* and *trans* configuration, respectively. The intensity ratio, I(*trans*)/I(*cis*) increases continuously with decreasing temperature. Similar changes in the intensity ratio were found in another pair of bands, 965 cm^{-1} (*cis*) and 976 cm^{-1} (*trans*), caused by the $\sigma(OH)$ mode. From these temperature dependent data, the enthalpy difference between the *cis* and *trans* configuration has been evaluated as 0.8 kJ/mol for stearic acid.

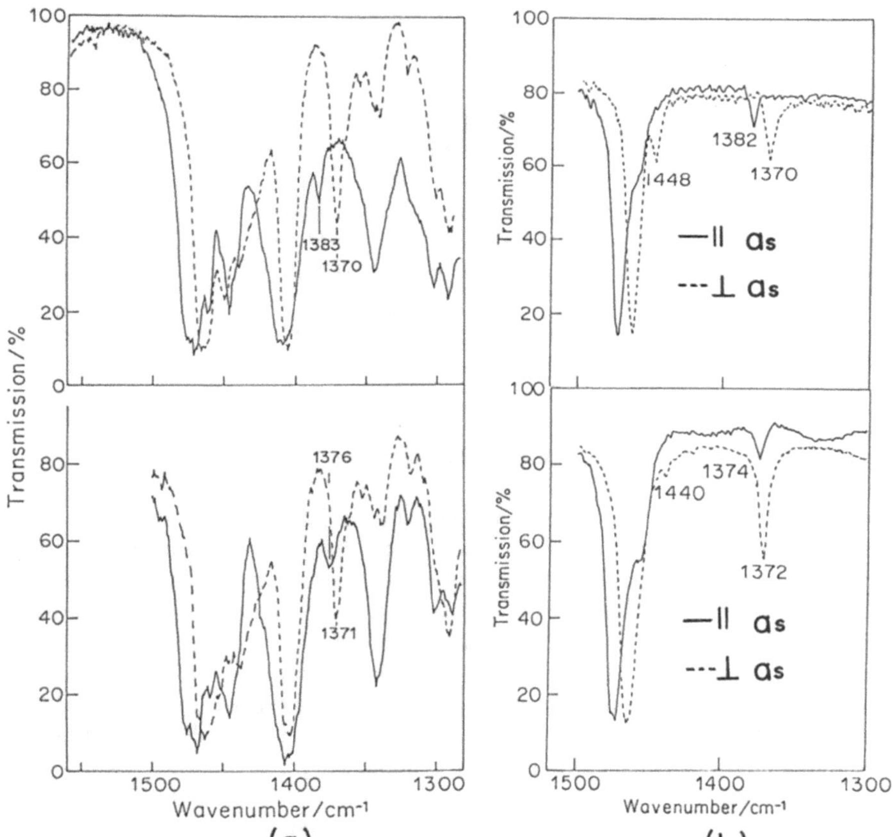

Fig. 10 a, b. Polarized infrared spectra of the orthorhombic (*upper part*) and of the monoclinic (*lower part*) polytypes of the B polymorph of stearic acid (**a**), and of hexatriacontane (**b**), at room temperature

The dynamic process of *cis-trans* conversion is thought to occur in conjunction with a simultaneous proton transfer.

In contrast, the δ(C=O) band in the B and in the E form of stearic acid appears as a singlet associated with the *cis* configuration, at 641 cm^{-1} for B, and at 684 cm^{-1} for E, at room temperature. This band did not show any appreciable change with temperature, indicating that only *cis* is permitted in the B and in the E form in the whole temperature range. The molecular motion in the C form is reflected in the Raman spectral pattern in the low frequency range. The spectra taken with a single crystal did not show a pronounced polarization dependence as would have been anticipated from the space group symmetry, and changed dramatically with temperature, in contrast to the case of B and E [48].

As to the sensitivity of Raman and infrared spectra to the polytypic structures, the following two spectral patterns are most characteristic. *First*, the different packing of the terminal methyl groups at the lamellar interface is reflected in the infrared absorption band, associated with the methyl deformation vibration. Figure 10 shows the polarized infrared spectra of two basic polytypes of (a) stearic acid [32] and (b) hexatriacontane [50]. In both cases, the two polytypes give rise to different positions of the absorption bands which are caused by the symmetric deformation of the methyl group, 1376 cm^{-1} (a_s-polarized) and 1371 cm^{-1} (b_s-polarized) in the B_m form, and 1383 cm^{-1} (a_s) and 1370 cm^{-1} (b_s) in B_o. The quite similar spectral patterns of the basic polytypes of stearic acid and hexatriacontane are ascribed to the similarity in the interfacial structures. *Second*, the Raman spectra in the very low-frequency region respond to the respective

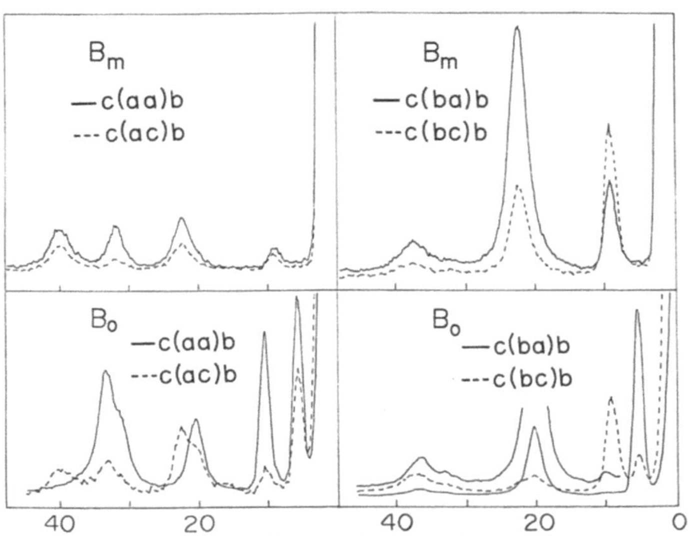

Fig. 11. Polarized low-frequency Raman spectra of the B_m and B_o polytypes of stearic acid at room temperature

polytypic structure. As shown in Fig. 11, the B_o and B_m forms of stearic acid reveal quite different Raman spectra in the region below 50 cm^{-1}. In this region appear the bands which are associated with the bending of the dimer chain as a whole, coupled with the rotation about the normal to the chain axis [32]. For B_m the (aa) and (ac), or (ab) and (bc) components appear at the same frequencies, whereas they split into different frequencies in B_o. The band splitting in the orthorhombic polytypes in this frequency range is caused by the lamella-lamella interaction between adjacent layers, as expected from group theoretical considerations [32, 50].

Obviously there is a systematic difference in the Raman frequencies (the optical phonons) between the two polytypes. The bands of B_m appear at frequencies higher than the midpoints of the corresponding doublets in B_o. This means that the density of states associated with the low-frequency optical branches in the B_o form is shifted toward lower frequencies than those in B_m, as shown in Fig. 11 [32]. The same tendency was obtained for the acoustic branches measured by Brillouin spectra [51]. The difference in the normal frequencies in the low-frequency region is closely related to the thermodynamic stability of the polytypic structures, as will be proved by the experiments in the next section.

3 Thermodynamic Stability of Polymorphs and Polytypes

It is clear that a precise knowledge of the thermodynamic stability is a prerequisite for an understanding of crystal growth of different polymorphs and polytypes. For this purpose, measurements of the solubility and of the melting point of each modification are most determinative, since the most stable form has the lowest solubility and, at the same time, the highest melting point [1]. Solid-state transformations may also give comprehensive data. However, these are sometimes influenced by kinetic effects due to certain steric hindrances. Until now, there are only a few substances among the long-chain compounds for which the stability of the different polymorphs and polytypes was determined: obviously stearic acid may be considered as the most representative one, as will be described below.

3.1 Polymorphic Stability

First the solubilities of the A, B and C forms of stearic acid were measured independently at temperatures ranging from −2 to 38°C [52]. (In this study, the distinction of B_o and B_m was disregarded.) The temperature at which the system, consisting of the solution and the single crystal, reached equilibrium, was determined for each polymorph and found to lie at different temperatures. The

Table 4. The enthalpy (ΔH_d) and entropy (ΔS_d) of dissolution of the A, B and C forms of stearic acid

Form	ΔH_d(kJ/mol)			ΔS_d(J/mol K)		
Solvent	A	B	C	A	B	C
Decane*	65.7	69.0	64.4	193.4	205.0	189.9
Methanol*	—	88.9	84.4	—	241.7	230.9

* The values at 27°C.

data of enthalpy and entropy of dissolution in two solvents are summarized in Table 4. It was confirmed by another experiment [53] that the B form has the lowest solubility below 32°C, while the lowest for the C form lies above that temperature. The solubility of the A form was always higher than the lowest values (of either B or C) at all temperatures examined.

To prove the growth of B at the expense of A and C in a nearly saturated solution below 32°C [54, 55], Fig. 12a shows the result in decane solution at 25°C [55]. The opposite result was obtained at 35°C [55], namely C grew at the expense of B and A. Consequently, the thermodynamic stability of A, B and C was unambiguously determined. In some earlier reports [56–58], C was thought

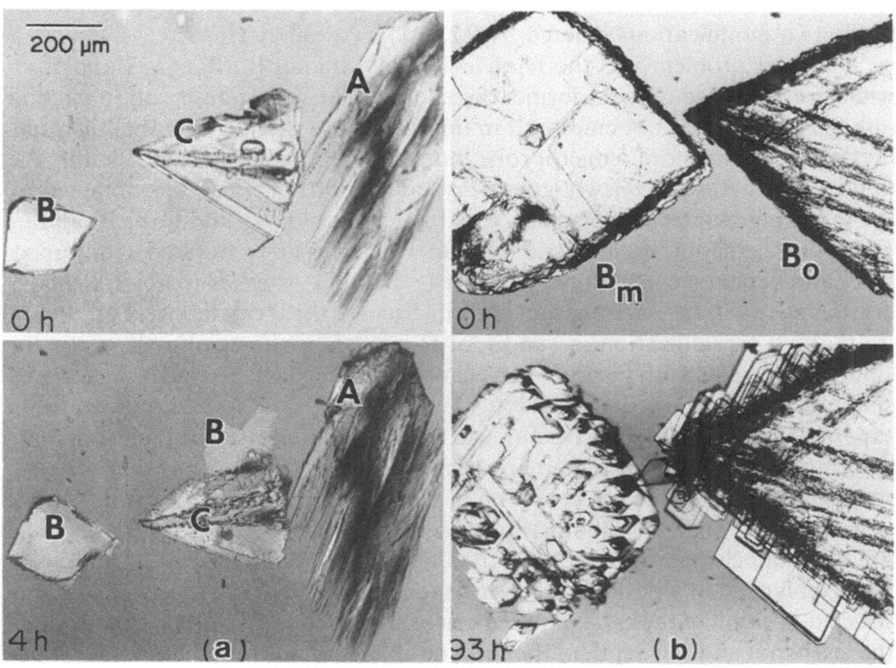

Fig. 12 a, b. Growth and dissolution processes of (**a**) the A, B and C polymorphs of stearic acid in decane at 25°C, and (**b**) B_m and B_o polytypes of stearic acid in decane at 30°C

most stable because, on heating, both A and B irreversibly transformed to C which then melted. This conclusion was misleading, since the kinetic effects in the solid-state transformation were disregarded. Bailey et al. [59] reported that the solubility became lowest for E, B and C with increasing temperature, but they did not measure the numerical values.

Finally, the crossing points of the solubilities of the B and C forms of palmitic, arachidic ($n_c = 20$), and behenic ($n_c = 22$) acid were found as 24, 40 and 46°C, respectively, below which B was always more stable [60].

3.2 Polytypic Stability

Experiments

The relative stability of the B_o and B_m forms of stearic acid was confirmed by observing the growth-dissolution process using a single crystal of each modification which was put into a nearly saturated solution [51, 61, 62]. The polytypic form was identified by low-frequency Raman data prior to the experiment [32]. Figure 12b shows that B_o has the lower solubility than B_m at 30.0°C. The same result was obtained at 20 and 15°C. From this it is clear that the B_m polytype is metastable in the range of temperatures examined. The saturation temperatures of the two modifications differed by 0.2 ± 0.1°C at 30 and 17°C.

The next problem was the relative stability among B_o, B_m, C, A and E, of which we examined only the former three forms. The simple observation method (used in Fig. 12) was not employed in this case, since B_m and B_o reveal the same crystal habit. Therefore, a microprobe Raman method was applied to identify the polytypic structure of a newly-grown crystalline portion which was obtained by the following two treatments [62]. First the crystals of B_m and C, or B_o and C, were put together in the *n*-hexane solution which was first saturated with respect to B_o and kept at a constant temperature (± 0.05°C). After several hours, the B_m crystals were withdrawn from the solution. Second, the B crystals were grown via *iso*-thermal crystallization; a supply of C crystals was put into the *n*-hexane solution saturated with respect to B_o below 32°C, then new B crystals crystallized spontaneously via the dissolution of the C crystals over several days. In both experiments, the polytypic structures in the whole interior of all B crystals were determined by microprobe Raman spectroscopy.

Figure 13 shows the optical set up of the microprobe Raman system [61] the which consists of an *epi*-illumination optical microscope (Olympus BH-2) and a double monochromator (Jasco CT-1000D). The incident beam (514.5 nm, Ar^+ laser), polarized in a desired direction, is focused through the objective lens at a selected position of the crystal placed on the microscope stage. The dimension of the laser spot and the depth of the focal point measured from the crystal surface can be adjusted to 1 μm and $\pm$ 1 μm, respectively, with an objective lens of $\times$ 100 magnification. The position subjected to the spectral measurement was moni-

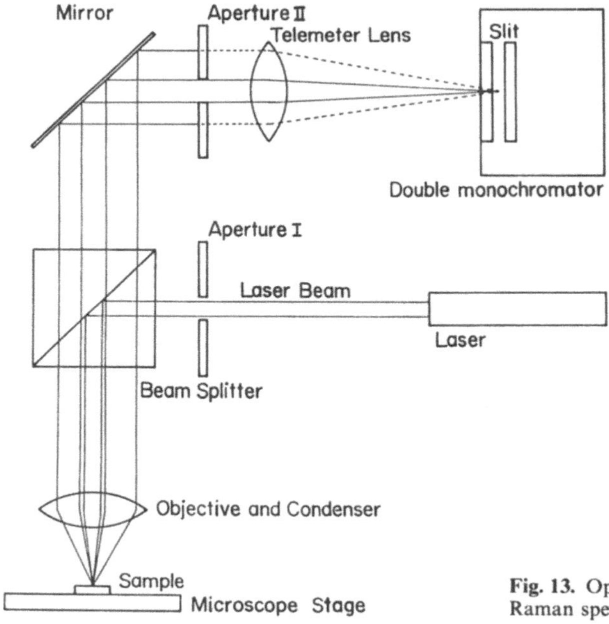

Fig. 13. Optical setup of the microprobe Raman spectroscope

tored by a TV camera. The backward scattered light was collected by the same objective lens, passed through a beam splitter, and entered the monochromator through a telescope. An analyzer and a polarization scrambler were put before the entrance slit of the monochromator as in the case of an ordinary Raman measurement. If necessary, an aperture of suitable diameter was inserted in front of the telescope in order to cut off the light passing through the peripheral area of the objective lens.

It was found that B_o has the lowest solubility below 30°C, since no conversion to other modifications was detectable in the overgrowth experiment. This is consistent with the results in Fig. 12, implying that B, whose solubility crosses that of the C form at 32°C, was of the orthorhombic polytype. Table 5 summarizes the results of the overgrowth experiments which subjected the B_m and C forms to iso-thermal crystallization. In the overgrowth experiment, the polytypic modification of the newly-overgrown crystal was converted to B_o above 22°C, caused by the dissolution of B_m in presence of excess C crystals, and the growth of B_o onto B_m. Below 22°C, no conversion from B_m was observed, since the solubility of B_m is lower than that of C. Figures 14a and b show the as-grown surface and microprobe Raman bands in the low-frequency region, respectively. The growth spiral steps are due to the overgrowth. The Raman bands of all portions at the surface covered with the spiral steps revealed the characteristic pattern of B_o. On moving the focal point from the surface (at point F in Fig. 14a) to the interior of the crystal, new bands, characteristic of B_m, appeared and increased in their intensities with a decrease in the bands of B_o. Ultimately, at the

Table 5. The overgrowth features and isothermal nucleation probability of the B_o and B_m polytypes of stearic acid from *n*-hexane solution [62]

T(°C)	Overgrowth on B_m	Isothermal crystallization*	
		B_O	B_m
30	—	35/35	0/35
28	B_o/B_m	25/25	0/25
25	B_o/B_m	—	—
24	—	13/13	0/13
23	B_o/B_m	9/21	12/21
22	B_m/B_m	11/19	8/19
21	B_m/B_m	—	—
20	—	3/8	5/8

*The denominators are the numbers of crystals examined.

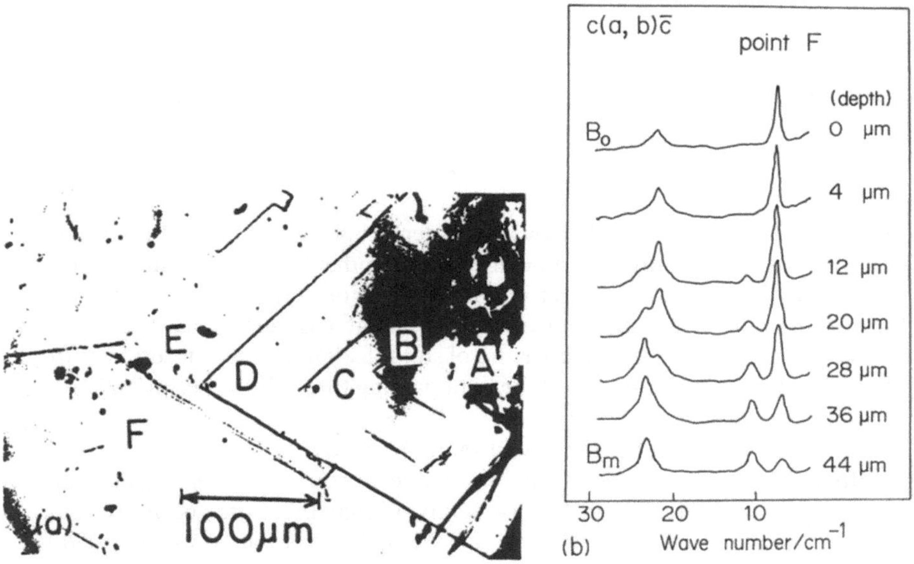

Fig. 14a. Spiral steps of the overgrown surface of a B_m seed crystal, and (**b**) microprobe Raman spectra taken at point F in part (**a**) of the figure at different depths, as indicated

depth of 40 μm, the Raman bands completely converted to those of the B_m crystal seed.

So far the isothermal crystallization yielded only B_o above 24°C, whereas both modifications crystallized below 23°C. Hence one may conclude that the temperature of the crossing point of the solubilities of B_m and C lies at 22 ~ 23°C from the data of the overgrowth, and at 23 ~ 24°C from the isothermal crystallization experiments. The small discrepancy may presumably be due to a small kinetic effect of crystallization involved in the above thermal treatments.

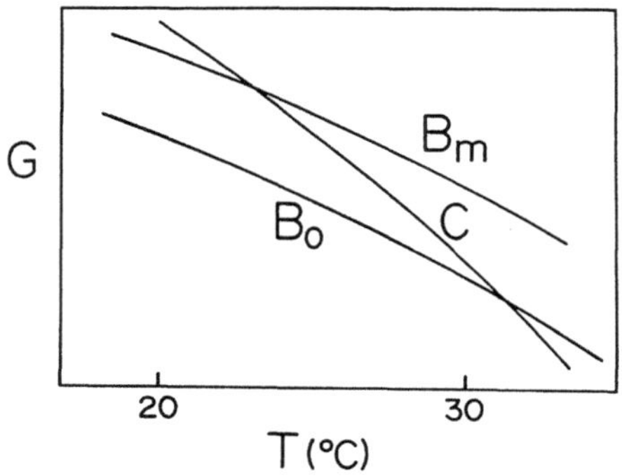

Fig. 15. A qualitative relationship of Gibbs free energy (G) and temperature (T) of the B_m, B_o and C forms of stearic acid

To summarize, the crystal Gibbs free energies of B_m, B_o and C are qualitatively drawn against temperature in Fig. 15 (G–T relationship). The A form is deleted here, since it was not examined in comparison with B_m and B_o; yet its G value is not the lowest at any temperature. There was no solubility data available of the E form. Despite the lack of data for these two modifications, the G–T relationship as shown in Fig. 15 seems to be the first case where the thermodynamic stability among the polymorphic as well as polytypic modifications has been determined in a group of long-chain aliphatic compounds. The experimental determination of the polytypic stability may be a rare case, since no stability field is expected for polytypism [63]. It is also worth noting, as shown in the results displayed in Fig. 15, that the basic monoclinic structure B_m of stearic acid is thermodynamically less stable than the pseudo-orthorhombic polytype B_o in the temperature range examined.

Spectroscopic Considerations

The molar Gibbs free energy of a crystal is expressed by the equation:

$$G = U + PV + A_v \tag{1}$$

where U denotes the lattice cohesive energy, P pressure, V molar volume, and A_v vibrational Helmholtz energy. In the harmonic approximation A_v is expressed in terms of the normal frequencies as:

$$A_v = (mN)^{-1} RT \sum_{j,q} \ln \{2 \sinh [h\nu_j(q)/2kT]\} \tag{2}$$

where $\nu_j(q)$ denotes the normal frequency of the j-th branch at the wave number q, m the number of molecules per unit cell, N the number of unit cells taken in the calculation, R the gas constant, k Boltzmann's constant, h Planck's constant. In an approximation neglecting dispersion, (i.e., if for every optical branch the normal frequencies at any wave number vector q are assumed to be the same as that at $q = 0$), the difference in A_v due to the optical phonons between the two polytypes can be evaluated from the observed zone-center frequencies. For the B form of stearic acid, $\Delta A_v = A_v(B_m) - A_v(B_o)$, was calculated as 0.73 kJ/mol at 30°C [50].

A contribution of the acoustic phonons to ΔA_v was estimated from the Brillouin scattering data [51]. The velocity v_s of the longitudinal hypersonic wave that propagates along the normal of the basal plane significantly differs between the two polytypes of B; $v_s = 4.212$ km/s in B_o, and 5.489 km/s in B_m at 30°C. The v_s values for the waves propagating within the basal plane are nearly the same in the two forms. The normal frequency of this particular acoustic branch at $q_c = 2.98 \times 10^6\,m^{-1}$, which corresponds to the scattering angle for the Brillouin measurement, is $0.410\,cm^{-1}$ in B_o and $0.547\,cm^{-1}$ in B_m. Assuming that these frequencies correspond to the Einstein frequencies of these polytypes, ΔA_v due to this acoustic mode is evaluated as 0.18 kJ/mol at 30°C. Thus the total ΔA_v is about 0.9 kJ/mol at 30°C. This value is approximately one tenth of that between different polymorphs.

The difference in U between the two polytypes was calculated from the interlamellar potentials [48], taking into account the terminal $-(CH_2)_6-CH_3$ groups at the lamellar interface. The van der Waals energy due to the non-bonded atom pairs was expressed by the sum of the pair-wise interatomic potentials, each term being given by the Buckingham function. The conformational energy of the terminal methyl groups was also calculated using the three minima sinusoidal potential function with a barrier height of 8.35 kJ/mol. The atomic positions within one layer were taken from the results of the X-ray analyses [14, 32]. The calculation, performed with varying lamellar repeating distance, L, and varying methyl conformation, resulted in almost the same minimum U values for B_o and B_m at $L = 43.8$ Å ($L = 43.86$ Å by X-ray data). Consequently, the difference in the calculated total Gibbs free energies of the two polytypes of the B form is due to A_v, which leads to a higher stability for the orthorhombic B_o form.

The C form of stearic acid always appears in the normal monoclinic form, and does not exhibit a pseudo-orthorhombic polytype like the B_o form. The calculation of the interfacial cohesive energies of C_m and a postulated C_o form showed that the U function for C_m has a narrow minimum at $L = 39.8$ Å ($L = 39.87$ Å by X-ray data), whereas C_o gave rise to rather shallow minimum at a larger L and, was destabilized by 2 kJ/mol compared to C_m. Thus the instability of C_o may be ascribed to the interfacial cohesive energy [68].

3.3 Solid-State Transformation

Numerous polymorphic transformations of the long-chain compounds are of first order, accompanied by a latent heat of transformation, and also with a thermal hysteresis.

3.3.1 Polymorphic Transformations

The typical polymorphic transformations in the *n*-alkanes may be classified into two kinds, in terms of the change in the subcell structure: orthorhombic-(pseudo-hexagonal)-hexagonal-melt, or triclinic-hexagonal-melt. The thermal data of numerous *n*-alkanes were summarized by Small [64].

As two representative examples characteristic for the *n*-fatty acids the transformations of stearic acid and of pentadecanoic acid are displayed in Table 6. All transformations in the stearic acid polymorphs are irreversible, and the transformed C crystal, particularly when originating from B, is destructed. Transformation in other even-numbered acids are similar. According to newer results [67], the transformation between B′ and C′ in pentadecanoic acid is reversible, although the A′–B′ transformation is irreversible. The reversible B′–C′, transformation was also observed in heptadecanoic acid, but A′ directly transformed to C′ in tridecanoic acid [56]. The transformation in the odd-numbered fatty acids will be described below. The irreversibility may be ascribed to large steric hindrance. In this regard, it is worth to note that the values of the temperature of transformation (T_{tr}) of B to C and of A to C for stearic acid are considerably higher than the crossing points of their G values: B_o–C at 32°C, B_m–C around 23°C, A–C around 35°C. This "overheating" may also be due to large steric hindrance, in which the rearrangements in the molecular conformation, the chain inclination with respect to the basal plane, as well as to the subcell axes are involved, as discussed in the next section.

Reversible polymorphic transformations in the C_{13}, C_{15} and C_{17} acids were recently studied by infrared spectroscopy [67]. Figure 16 shows the infrared spectral change during the B′–C′ transformation in pentadecanoic acid. During the transformation, the intensities and the positions of the bands originating from the *all-trans* hydrocarbon chains remain almost unaltered, except for a little reduction in the band gaps of the $\delta(CH_2)$ and $r(CH_2)$ doublets. This indicates a depression of the interchain forces in the C′ form. Conversely, the bands due to the carboxyl groups, i.e., the $\nu(C{=}O)$, $\sigma(OH)$ and $\delta(C{=}O)$ bands revealed

Table 6. Polymorphic transformations in stearic acid and pentadecanoic acid which occur at the temperature T_{tr}(°C), with the transformation enthalpies ΔH_{tr}(kJ/mol) indicated

	Stearic acid			Pentadecanoic acid[a]	
	A → C[b]	B* → C[c]	E → C[c]	A′ → B′	B′ → C′
T_{tr}	64	51	43.5	44	45.6
H_{tr}	very small	5.1	n.a.**	0.7	5.7

* The B_0 and B_m polytypes are not separated.
** Data are not available.
[a] Ref. [67]
[b] Ref. [65]
[c] Ref. [66]

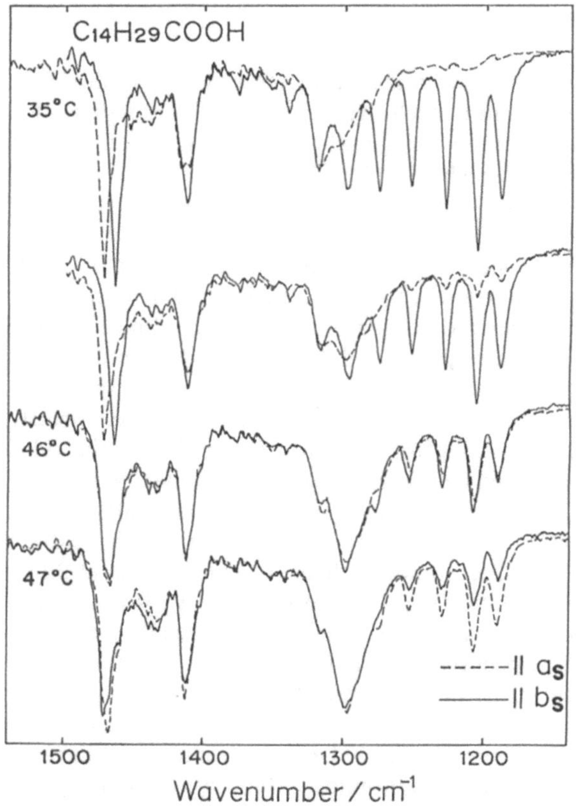

Fig. 16. Infrared spectral changes during the B′–C′ phase transformation of pentadecanoic acid

remarkable changes. This suggests that the transformation is associated with a reconstruction of the polar groups, rather than a conformational change in the hydrocarbon chain. Furthermore, the polarization of the $\delta(CH_2)$ and $r(CH_2)$ doublets, as well as that of the ν_3 progressive bands, exhibited very characteristic changes. On heating, the polarization first diminished, and the $\delta(CH_2)$ and $r(CH_2)$ doublets changed to broad singlets immediately below the transformation temperature (46°C). At 47°C, a clear but opposite polarization was recovered. This spectral change was followed reversibly and reproducibly during cyclic temperature changes, indicating that a reversible 90° rotation of the $O_\perp$ subcell took place between B′ and C′ by passing through a hexagonal packing (Fig. 17a). In case of A′–C′ of tridecanoic acid, it was also confirmed, by a similar experiment as that of Fig. 16, that the change in the subcell from $T_\parallel$ to $O_\perp$ is caused by a collective rotational movement of the chains, as illustrated in Fig. 17b.

Finally no polytypic transformation between B_m and B_o was observed in the solid state, below the temperatures at which both polytypes of B transformed to

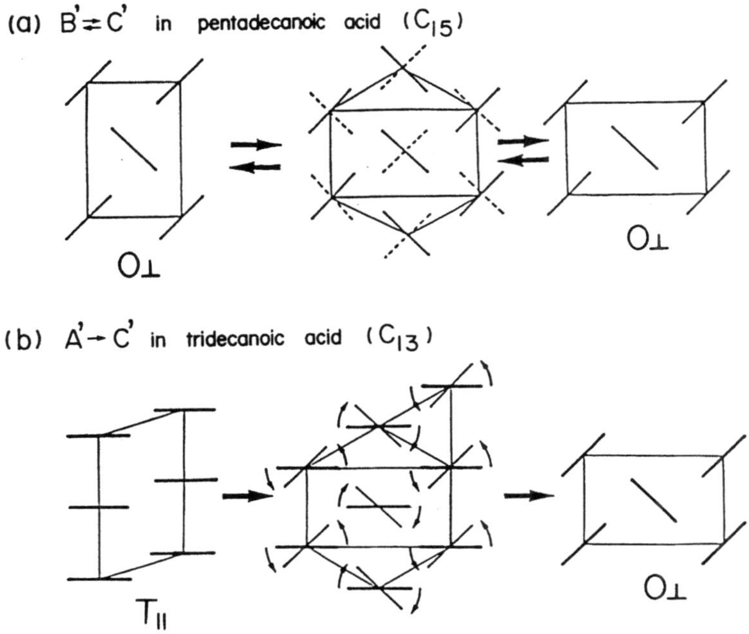

Fig. 17 a, b. Change in the subcell dimensions during (**a**) B′–C′ transformation of pentadecanoic acid and (**b**) A′–C′ transformation of tridecanoic acid

C. It was also found that no appreciable change was detectable in T_{tr} and in the transformation enthalpy from B_o to C and from B_m to C [66].

3.3.2 Composite Polymorphic/Polytypic Transformations

It is interesting to know how the polytypic structure is influenced during a change in the polymorphic structure. This could be examined for B_o, B_m and C of stearic acid in which the conversions in the polymorphic and polytypic structures occurred simultaneously [68] (composite polymorphic/polytypic transformation). This transformation was easily examined by observing the morphology of mono-lamellar steps with electron microscopy, because the B–C transition is often destructive.

Figure 18 shows the electron micrographs of germanium-decorated replica of the partially evaporated (001) surfaces: (a), (b) and (c) of the single crystals of B_m, B_o and C, and (d) and (e) of C transformed from B_m and B_o, respectively. The evaporation steps originate from terminal ends of edge dislocations in (a) and (b) and of screw dislocations in (c), (d) and (e). The lines of all these dislocations are directed normal to the (001) faces. The details of the appearance of evaporation steps due to lattice defects in stearic acid were precisely studied elsewhere [69].

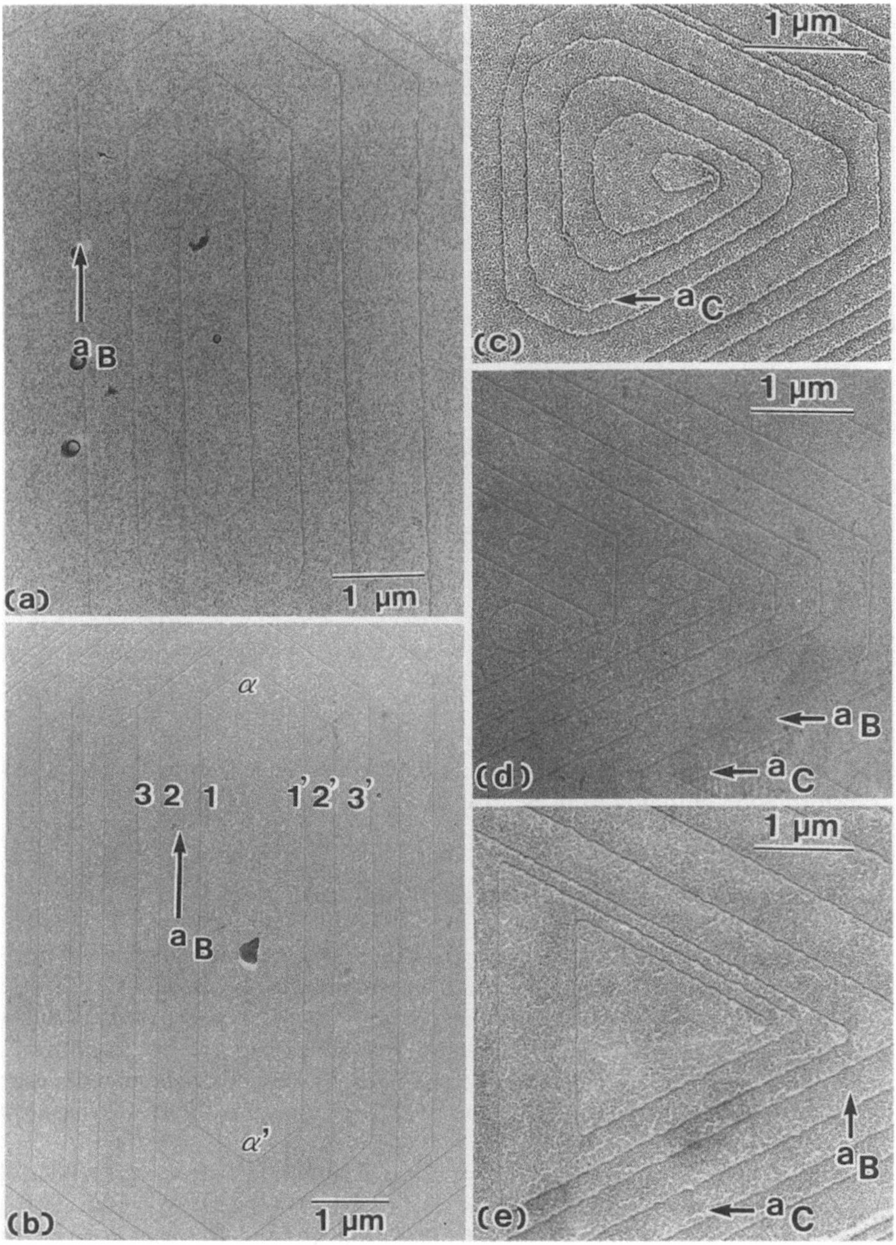

Fig. 18 a–e. Stearic acid, monolamellar evaporation steps of (**a**) B_m, (**b**) B_o, (**c**) C transformed from B_m (Fig. 18a) and (**e**) C transformed from B_o (Fig. 18b)

In B_m and B_o, the step patterns are of hexagonal shape consisting of [110] and [100] steps. The step pattern of B_m is compatible with a monoclinic symmetry of $P2_1/a$. The step distances along the b axis are the same, whereas they are different between those propagating upward and downward along the a-axis. This difference originates from the tilt of the long-chain axis. In B_o, an interlaced pattern is revealed, being characteristic of the double-layered polytype, as revealed in the earlier studies with an optical microscope [33, 70, 71]. The [100] segments of the steps of 1–1′, 2–2′, 3–3′, etc. have more or less the same propagation velocity. The 1–1′ step propagating toward the α direction along the a-axis reaches the α' portion of the 2–2′ step. In contrast, the α' portion of the 1–1′ step does not reach the 2–2′ step along this direction, since each successive layer differs alternately in the tilt direction of the long-chain axis. The same mechanism accounts for each couple of successive layers in B_o, causing the interlaced step pattern. As for C, the step pattern was also compatible with $P2_1/a$. No interlaced pattern like B_o was observed, indicating exclusive occurrence of the single-layered modification of C in stearic acid.

The electron micrographs of C which transformed from B_o and B_m at 62°C (Fig. 18d and e), first showed that the destructed C crystals were actually composed of many small grains whose dimensions ranged from 10 to 100 μm. Each grain retained single crystallinity, since the regular step morphology of C is revealed on each grain surface. This means that a certain collective lattice displacement should have occurred during the transformation. The destruction through the transformation may primarily be due to a reduction of the lamellar thickness (Table 3), causing a slip of the molecules along the long-chain axis at the grain boundary. Second, all step patterns of the transformed C crystals were equidistant, which is characteristic of the single-layered structure. This means that even the double-layered structure of B_o was not preserved through the transformation. Furthermore, when the a-axis of the C crystal (a_C) was compared to a_B, by observing the shadowing direction of germanium (which was always along the a_B direction) then, besides random orientations, two coherent relations were confirmed: $a_B \parallel a_C$, Fig. 18d, and $a_B \perp a_C$, Fig. 18e. The former relation appeared in the $B_m \rightarrow C$ transformation (reproduced for 10 crystals). In most recent experiments, the $B_o \rightarrow C$ transformation predominantly resulted in $a_B \perp a_C$ (reproduced for 20 out of 23 crystals), and in fewer cases in $a_B \parallel a_C$ (reproduced for 3 out of 23 crystals), although the latter case was not reported in [68].

According to the structure data (Table 3), one may interpret that the $B_m \rightarrow C$ transformation characterized by $a_B \parallel a_C$ involves an increase in the chain tilt and a displacement of the a_s and b_s axes of the $O_\perp$ subcell. In turn, the $B_o \rightarrow C$ transformation involves a rotational displacement of the symmetry axis whose direction is opposite in the two adjacent lamellae, so that the single-layered C is formed. In case of the relation of $a_B \perp a_C$, displacements of the a_s and b_s axes are also included. Both mechanisms are illustrated in Fig. 19. Consequently, one may perceive that the steric hindrance in the $B \rightarrow C$ transformation involves specific lattice displacements. These displacements may be characteristic for the n-fatty

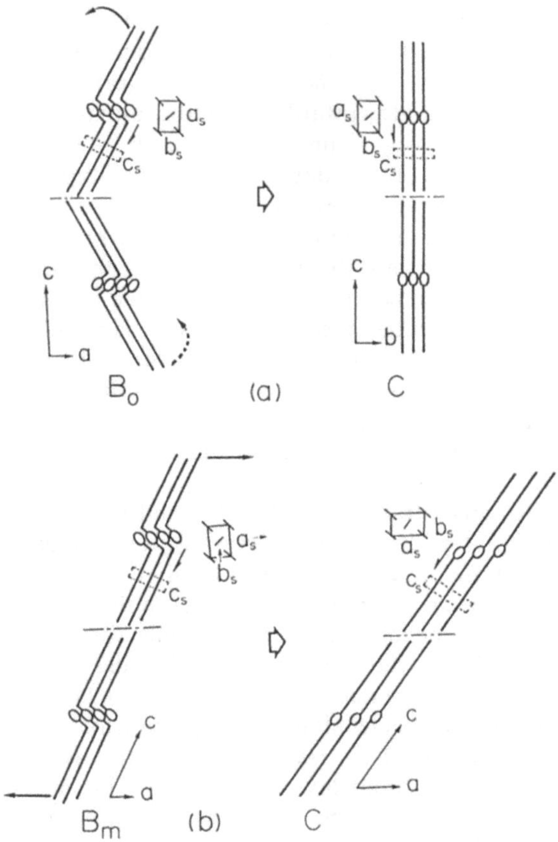

Fig. 19. Schematic illustrations of the lattice displacements in the transformations of $B_o \rightarrow C$ (a), and $B_m \rightarrow C$ (b), of stearic acid

acids, since they occurred reversibly in pentadecanoic acid (Fig. 17) and, irreversibly in the $E \rightarrow C$ transformation of stearic acid, a fact which was quite recently confirmed [72].

4 Crystal Growth

4.1 Solution Growth

In the past several years, crystal growth of different polymorphs from solution has been elucidated to a greater extent, using stearic acid as a model substance. The crystal growth of different polytypes, however, has been rather disregarded. This was because the observation of the crystal habit, which is the same for the different polytypes, was the only tool for an identification of the polymorphs in the usual crystal growth experiments. Therefore, in the crystal growth of stearic acid, $\mathbf{B}_o$ and $\mathbf{B}_m$ were not distinguished expcept for a few cases, although they should often have coexisted under usual growth conditions.

Occurrence

The occurrence of the A, B and C forms of stearic acid was examined in cyclohexane [54], methanol, benzene and decane [55] solutions. Table 7 shows data for decane and methanol [55]. The occurrence probability of each polymorph was evaluated for the first appearing crystals at different temperatures and initial supersaturations. The general trend was the following: in decane A and C prevailed at higher supersaturation, whereas B preferably occurred at lower supersaturations. In methanol, the occurrence of B was enhanced in comparison to decane used as a solvent. In both solvents, the relative abundance of A and C increased with increasing temperature and, consequently, that of B decreased.

The occurrence probability is primarily governed by the nucleation process; the successive growth may be less determinative. Hence, the nucleation function of B and C at 17°C obtained in decane solutions is drawn in Fig. 20. A simple nucleation function, J, can be written as [73]:

$$J = C_o \exp\{-[C'\Omega^2\gamma^3]/[(kT)^3(\ln\beta)^2]\} \tag{3}$$

in which C_o is a kinetic parameter, C' a shape factor of a considered crystal nucleus, Ω the molecular volume, k Boltzmann's constant, T temperature, γ a

Table 7. The occurrence probability of the A, B and C forms of stearic acid in decane and methanol solutions at different crystallization temperatures (T_c) and at different initial supersaturation (σ)

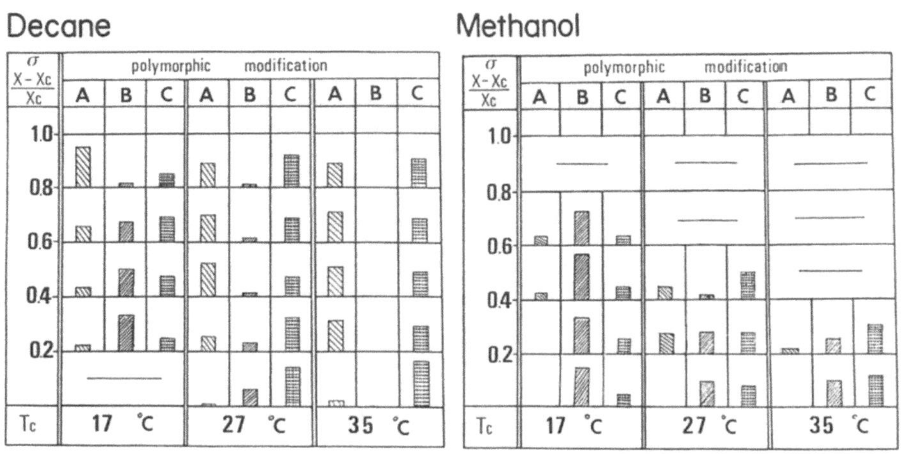

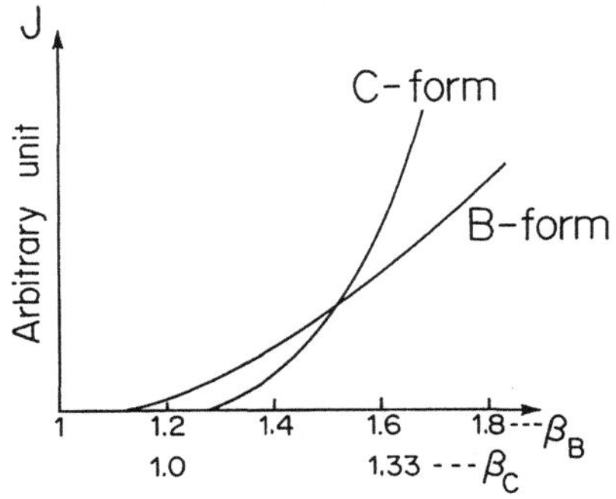

Fig. 20. Nucleation functions (J) of B and C in decane at 17°C, derived from data in Table 7

crystal-solution interfacial energy, and β the supersaturation ratio ($\beta = X/X_s$ where X is the molar fraction of the actual solution and X_s is the molar fraction in the saturated solution).

According to the difference in the solubilities of B and C, one obtains the difference in the β values between B and C as ~0.2 at 17°C. Therefore, J_B must be larger than J_C at lower β values. At larger β values, however, J_C exceeds J_B due to the smaller γ value for C. This is supported by the calculation of the surface energy in a crystal-vapor system of B and C [74]: the surface energies of all important faces, (001), (110), (100) and (010), are smaller in C than in B. One may reasonably perceive that this tendency may persist in the crystal-solution interfacial energy in decane, because stearic acid and decane form a nearly ideal solution [52, 75]. At 35°C, the occurrence of C in decane is always manifested. Both factors of β and γ dominate for the C form at this temperature, since C is the most stable form.

The results in decane may be discussed in terms of Ostwald's law of stages, which predicts that a metastable state tends to appear first under non-equilibrium conditions. Recent considerations on crystalline polymorphism showed [76] that the polymorph with the higher probability of occurrence can be a different one at different supersaturations, and depends not only on the surface energy, but also on bulk parameters of the solution. As a result, the preferential

Table 8. Relative occurrence of the B_o and B_m polytypes of stearic acid from *n*-hexane solution. T_s(°C) and T_c(°C) are saturation and crystallization temperatures, respectively. The denominators specify the total numbers of crystals evaluated at the respective temperature

T_s	T_c	B_m	B_o	Intermediate*
14	11	14/15	1/15	—
15	13	18/23	5/23	—
16	14	5/5	0/5	—
18	16	16/19	3/19	—
23	21	11/28	10/28	7/28
24	22	13/24	9/24	2/24

* "Intermediate" means crystals revealing diffuse Raman bands which differ both from B_o and B_m.

occurrence of the metastable polymorph can be associated with a narrow range of supersaturation, as seen in an example of an azo disperse dye [77], and in the case presented here.

In methanol solutions, B was always preferred at 17 and 35°C. This indicates the effect of the polarity of solvent. Phenomenologically, the preferential occurrence of B in polar solvents is consistent with previous data [57, 78].

Finally, the occurrence of B_o and B_m from *n*-hexane solution was examined for crystals grown at different temperatures with more or less the same initial supersaturations [62] (Table 8). All crystals were examined by low-frequency Raman spectroscopy. There was a subtle tendency that B_m crystallized more frequently than B_o with decreasing temperature.

Solution-Mediated Polymorphic Transformation

Solvent effects on the crystal growth of the polymorphic modifications of stearic acid clearly appeared in a solution-mediated transformation. This transformation is a result of concurrent processes of dissolution of the less stable forms and growth of the more stable ones in nearly saturated solution [79].

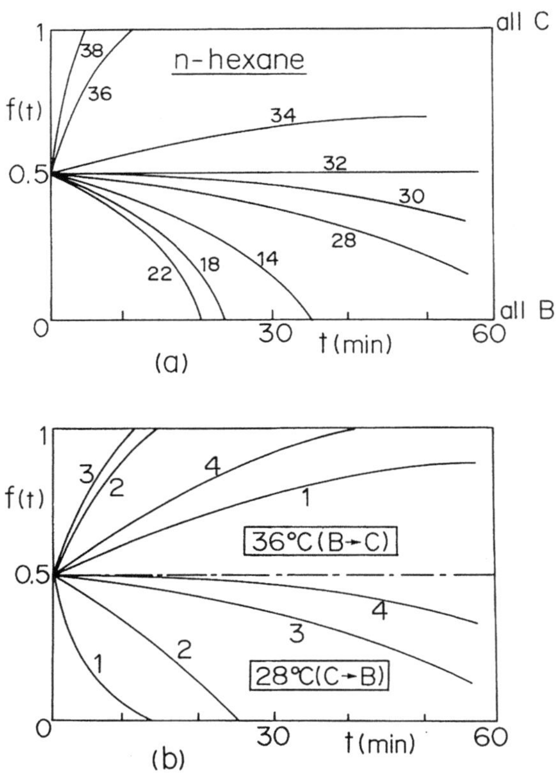

Fig. 21 a, b. Rates of solution-mediated transformations between the B and C forms of stearic acid; (**a**) in *n*-hexane for the temperature parameters indicated, (**b**) solvent effects, *1*: methanol, *2*: butanone, *3*: *n*-hexane, and *4*: decane

Figure 21 shows the rates of transformation, which are expressed as the rate of change in the fractional concentration of the C form, f(t), present in the crystal powders suspended in the solution [53]. The initial value of f(t) was 0.5, since the same amount of B and C crystals were put into the solution which was saturated with respect to the more stable form at the temperature examined. The data in the *n*-hexane solution (Fig. 21a) show that the transformations occurred from C to B below 32°C, and in the reverse direction above that temperature. In comparison with the solubility experiments, this is quite consistent. For the C → B transformation, the rate first increased, then decreased with increasing temperature. The maximum rate appeared at 22°C. This change is a result of two competing factors; a driving force for the transformation due to the difference in solubility, and the kinetics of growth of B and dissolution of C. On approaching 32°C, the former decreased, whereas the latter increased. Under the assumption that the transformation from C to B in *n*-hexane is growth-controlled [79], one may describe the transformation rate df(t)/dt as:

$$-\mathrm{df(t)/dt} = \mathrm{d[B]/dt} = \mathrm{k'[B]} \tag{4}$$

where [B] is the relative number of B crystals at time t divided by the initial number. The parameter k′ is the growth rate of B, being dependent both on supersaturation (σ) and on temperature:

$$\mathrm{k'} = \mathrm{b}\exp\{-[\Delta E_g(\mathrm{B})]/[\mathrm{RT}]\}\sigma^2 \tag{5}$$

In Eq. (5), b (time^{-1}) is a kinetic factor and ΔE_g(B) is the activation enthalpy for growth of B, R is the gas constant. Use of σ^2 assumes that the growth of B is by a spiral growth mechanism. σ can be calculated from the solubility data. The curves with best fitting to f(t) drawn in Fig. 21a yielded $b = 4 \times 10^{27}$ and ΔE_g(B) = 150 kJ/mol [53].

The solvent effect is better understood in Fig. 21b, examined in four solvents at 28 and 36°C. The transition rate for C → B increased in the following sequence:

decane < *n*-hexane ≪ butanone < methanol

whereas for B → C at 35°C in the sequence:

methanol < decane < butanone ≪ *n*-hexane.

The C → B transition occurred most rapidly in methanol. Among the predominant factors concerning the solvent effects, the solubility and viscosity effects were less pronounced in methanol [53]. Hence a microscopic origin was discussed [80], taking into account the facts that monomers are favored in polar solvents whereas dimers are favored in the nonpolar ones, particularly in short *n*-alkanes [52, 75, 81, 82]. The difference in the molecular structures of the (110) faces, which play a dominant role in the growth, was also considered. The monomers of stearic acid in methanol solution may be more efficiently adsorbed on the twisted surfaces of B, due to the C_2–C_3 *gauche* conformation, than on the surface of the C form which is flat due to the all *trans* conformation. This consideration may agree with the fact that the growth of B in polar solvents was depressed more than C by addition of surfactant molecules, as proved by the growth rate measurement in butanone [83], and by the solution-mediated transformation in methanol [84].

The ΔE_g value obtained from Eq. (5), 150 kJ/mol in *n*-hexane, may be compared with those values which were obtained from the growth rate experiments with single crystal, obeying the following relation:

$$V = p\sigma^n \tag{6}$$

$$\partial \ln p/\partial(1/T) = -\Delta E_g/R \tag{7}$$

In Eq. (6), n = 2 for spiral growth mechanism, as verified in *n*-alkane solvents [85], yielding the values of 150 kJ/mol and 170 kJ/mol for the (110) and (001) planes, respectively, which may justify the value obtained from the transformation data.

The intensive work on the linear growth experiments of B and C of stearic acid available [83, 85–87] was recently reviewed by Beckmann et al. [88]. It is worth noting here that the ΔE_g values of B are larger than those of C, in butanone and 2-octanone, which are both polar solvents, but, conversely, smaller than those of C in decane [89]. The differences are 73 kJ/mol in butanone [83] and 85 kJ/mol in decane [89], when the data are fitted to a parabolic V-σ relation. This contrast may be a critical measure to elucidate the solvent effect. The importance of surface diffusion processes involving step diffusion has been considered [83], yet the microscopic interpretation taking into account the growth units and stereochemical effects of the polymorphic structure may still be open to questions. Furthermore, the role played by the different polytypic structures must be examined more precisely.

4.2 Vapor Growth

Growth of thin films of long-chain compounds has attracted new attention. One of the aims may be to utilize the thin films in functional organic films. In doing so, two methods have been applied; Langmuir-Blodgett (LB) multilayer deposition [90] and physical vapor deposition [91]. Under the necessity of fabricating devices in a dry system preference is given to the latter method, although molecular orientation can be well controlled in the former technique. Microscopic observations of LB films have indicated that the uniformity of the molecular packing was incomplete, even in the first adsorbed monolayer [92, 93].

On the other hand, vapor growth has difficulties with arranging long-chain molecules in such a well-defined way as the LB method can do. Hence, it is important to investigate the molecular orientation and packing in the vapor-grown thin films of long-chain compounds. This section tries to briefly review recent work on the physical-vapor-deposited thin films of *n*-alkanes, *n*-fatty acids and their metal salts.

Figure 22 shows an electron micrograph of a germanium-replica of a stearic acid film deposited on an air-cleaved face of muscovite at 30°C in a vacuum of 10^{-6} Torr with a film thickness of 1000 Å [94]. Two kinds of growth hillocks are detectable; a normal growth hillock (NGH) in which the stearic acid molecules are arranged normal to the substrate, and a lateral growth hillock (LGH) with the molecules parallel to the substrate. The polymorphic modifications of the deposited stearic acid films were found to be the C form from the inter-step angle (55°) [94] on the NGH, and also by X-ray diffraction [95].

The molecular orientation in each growth hillock was determined by two techniques. First, the patterns of germanium decoration were quite different on the basal (001) faces, (island-like), and on the lateral (110) faces, (continuous), as examined using a large single crystal of stearic acid [94]. This difference was reproducibly confirmed, when the average thickness of the decorated germanium was below 50 Å. The second technique was to directly take the electron diffraction (ED) patterns of the growth hillock with a high-resolution electron microscope. Figure 23 shows the electron micrograph and the ED

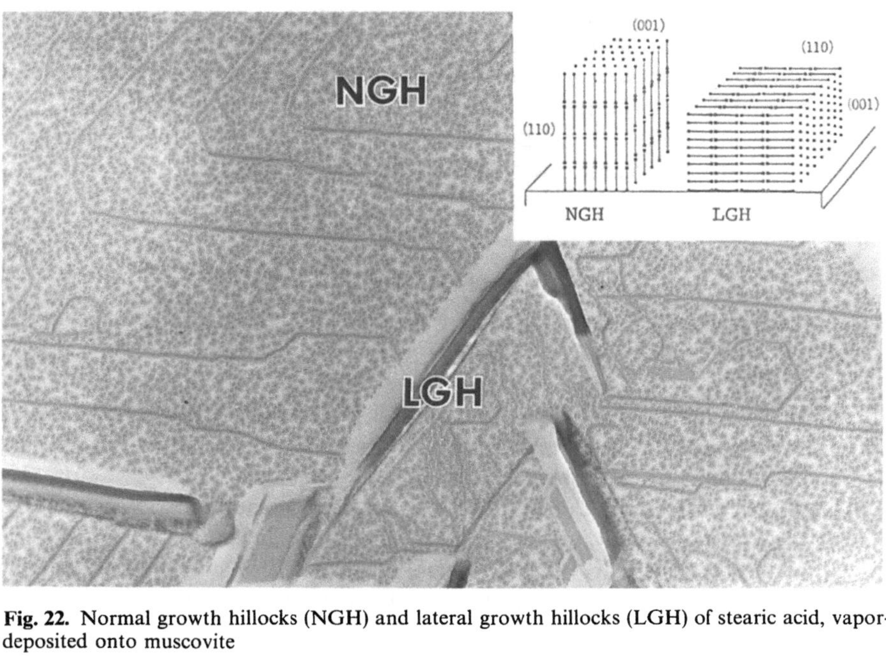

Fig. 22. Normal growth hillocks (NGH) and lateral growth hillocks (LGH) of stearic acid, vapor-deposited onto muscovite

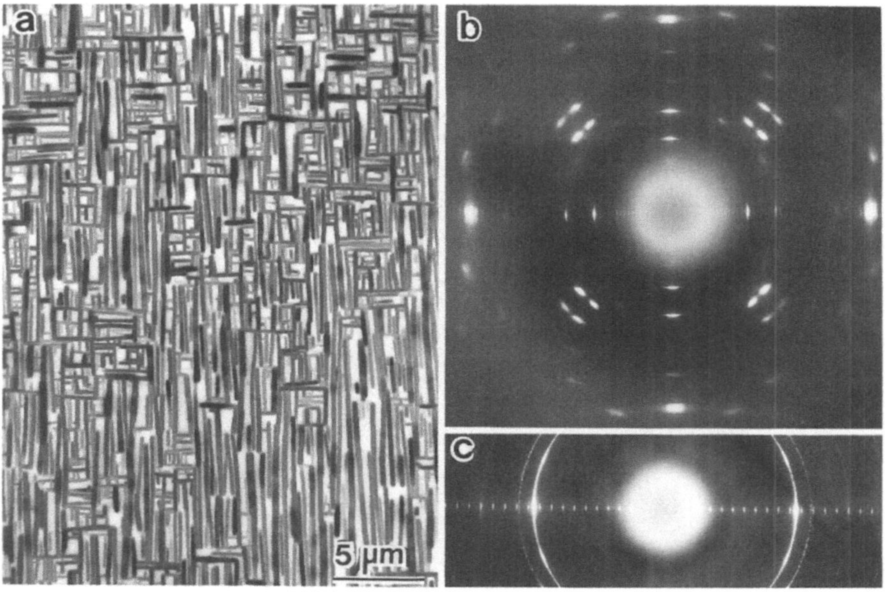

Fig. 23 a. Electron micrograph, and (**b**) and (**c**) electron diffraction patterns of thin films of hexatriacontane, vapor-deposited onto KCl

patterns of a LGH of hexatriacontane deposited on (001) of KCl [96]. The substrate was annealed at 150°C in vacuum after cleavage in air, with subsequent deposition at room temperature. The needle axes of the LGH are parallel to [110] of the KCl substrate, exhibiting an epitaxial relation (Fig. 23a). The ED patterns (Fig. 23b and c) consist of the diffraction spots of the $O_\perp$ subcell, $a_s = 7.42$ Å, $b_s = 4.954$ Å, $c_s = 2.546$ Å, and also of the (001) reflection of $d_{001} = 47.7$ Å. Similar ED patterns were obtained with films of methyl tricontanoate [97], *n*-alkanes on KCl having $n_c = 30 \sim 44$ [98], and calcium stearate [99]. As to the NGHs, Fig. 24 shows (a) a germanium-replica image of calcium stearate [99], and (b) an ED pattern of cadmium arachidate [100]. Both films were deposited on a cleaved face of muscovite at 70°C in a vacuum of 10^{-6} Torr. The growth spiral in Fig. 24a means that this NGH was initiated from the terminal end of a screw dislocation. The ED pattern yielded the parameters of the unit cell, $a = 4.9$ Å and $b = 7.6$ Å [100].

To relate the molecular orientation with the growth conditions, the effects of temperature and of the number of CH_2-units, n_c, were examined for the *n*-fatty acids. Figure 25 shows the fraction of the film area occupied with the LGH, C(LGH). Fatty acids having $n_c = 16 \sim 22$ were used [101]. It is clear that C(LGH) decreased for each acid with increasing temperature; i.e. the normally-oriented growth hillocks increased at the expense of the lateral ones. The effect of n_c also manifested itself in that C(LGH) increased when n_c became large. This means that

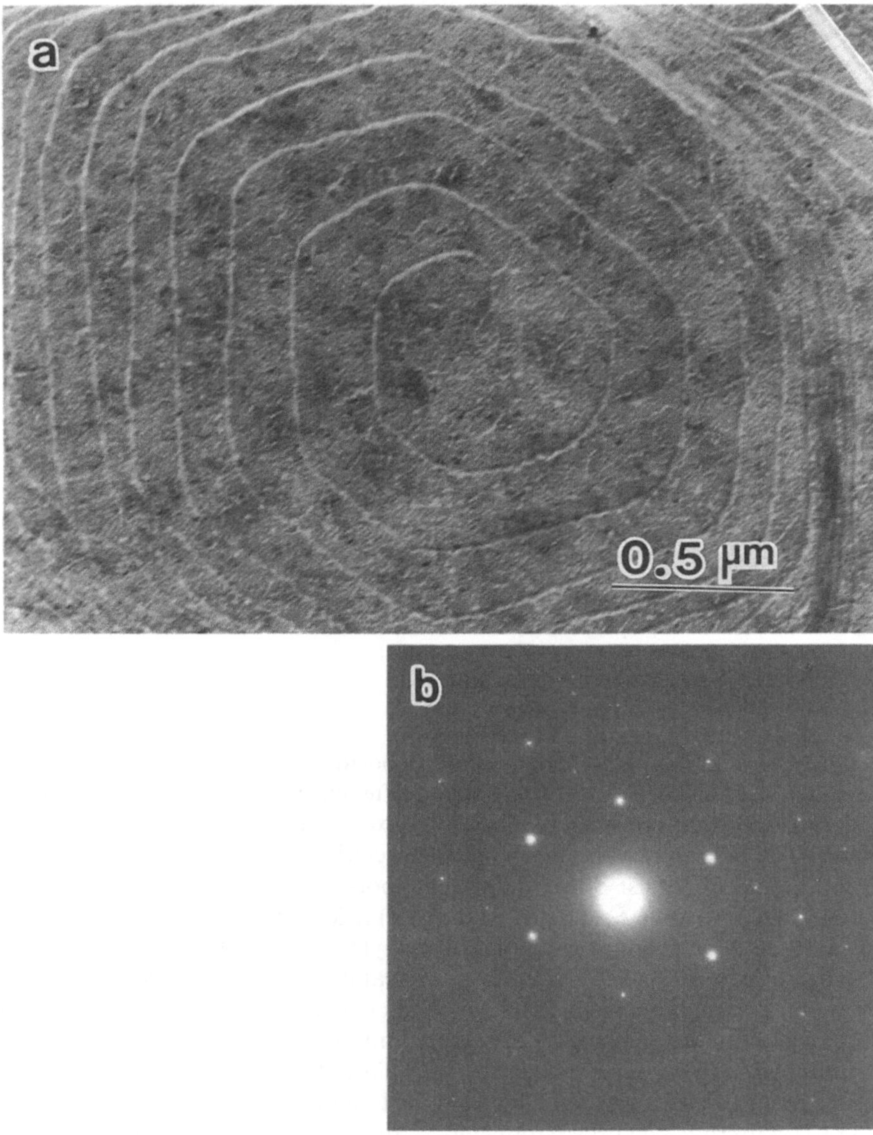

Fig. 24 a. Electron micrograph of a replica image of a normal growth hillock of calcium stearate, and (**b**) electron diffraction pattern of a normal growth hillock of cadmium arachidate; both materials were vapor-deposited onto muscovite

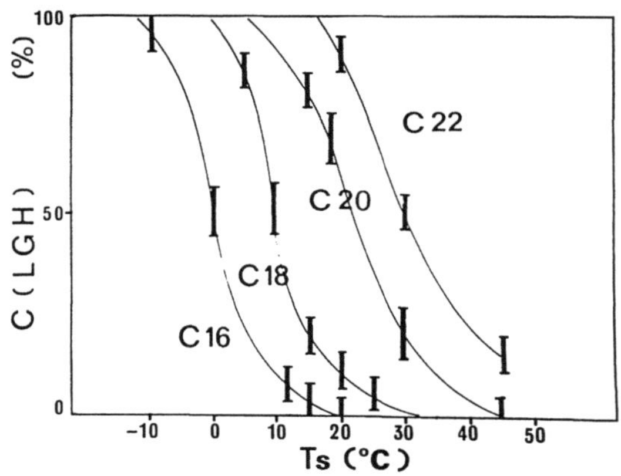

Fig. 25. Temperature dependence of the concentration of lateral growth hillocks C(LGH), in thin films of palmitic acid (C_{16}), stearic acid (C_{18}), arachidic acid (C_{20}), and behenic acid (C_{22}), vapor-deposited onto KCl

the lateral orientation is governed by the lateral interactions between the molecules and the substrate. It must be noted that the effect of the temperature of the substrate may involve both thermal and kinetic factors. To discriminate between these two effects, C(LGH) was measured by changing the molecular flux rate at a constant temperature [94]. The results showed that, at each temperature, C(LGH) increased nonlinearly with increasing flux rate. Due to a remarkable increase in the vapor pressure with raising temperature, the actual supersaturation for the vapor growth drastically decreased, as calculated for stearic acid [94, 102]. Consequently, the apparent effect of temperature in Fig. 25 was, in fact, primarily due to a kinetic effect of supersaturation. This conclusion leads to the idea that continuous thin films of stearic acid lying normal to the substrate may optimally be formed under lower actual supersaturations, and vice versa. This may apply to other long-chain compounds as well.

5 Summary

Great progress has recently been made in determining the molecular structures of the polymorphs and polytypes of long-chain aliphatic compounds. Accordingly, thorough information on thermodynamic as well as on kinetic aspects has been obtained on the transformation and crystal growth processes of this class of materials. This progress is based on the application of new techniques, particularly, of modern methods of vibrational spectroscopic analysis.

For further clarification, investigation of the microscopic mechanisms of the polymorph- or polytype-dependent nucleation and crystal growth processes will definitely be important. So far, more phenomenologically oriented approaches have been made in that only the rates of occurrence and growth of the different forms have been studied. In these studies, relations between the nature of the solvent and the kind of polymorphism occurring were found. A consideration on the role of the activation enthalpy for growth may lead to some ideas about the most critical activation steps, as has been demonstrated for *n*-alkanes [2] and for *n*-fatty acids [83]. In doing so, structural information on a molecular scale at the crystal-solution interfaces as well as about the adsorbants will be a prerequisite, possibly employing microprobe techniques of vibrational spectroscopy.

As to new classes of substances, long-chain compounds which are related to biologically active systems, such as phospholipids [103], acylglycerols and unsaturated fatty acids [104], will certainly attract more attention in the future. So far little is known about these compounds, in contrast to the *n*-alkanes and the saturated fatty acids. One may anticipate that molecular information about their polymorphism can be used to examine the critical roles which these biomembrane-constructing long-chain compounds play in biological activities.

Finally the thermodynamic and kinetic features of polytypism in the relevant substances also promise to become a subject of vital interest. In the long-chain compounds, the nature of polytypism is closely related to the lamella-lamella interaction, which can be separated from the lateral interactions due to a strong anisotropy in the molecular interactions. In this case, the vibrational spectroscopic techniques can provide very detailed data, as displayed in some examples in this article. This latter method can be useful especially for *in-situ* analysis of the crystal growth processes of the polytypes, which have a more complicated crystallization behavior.

Acknowledgements: The authors are indebted to Professor K. Inaoka, Dr. K. Yase and Dr. Y. Ueda for providing original micrographs.

6 References

1. Verma AR, Krishna P (1966) Polymorphism and polytypism in crystals, Wiley, New York
2. Boistelle R (1980) Defect structures and growth mechanisms of long-chain alkanes, in: Current Topics in Materials Sciences, Kaldis E (ed.), North Holland Publ. Co, Amsterdam, 413
3. Vand V (1951) Acta Cryst. 4: 104
4. Abrahamsson S, Dahlen B, Lofgren H, Pascher H (1978) Prog Chem Fats Other Lipids 16: 125
5. Hernqvist L (1988) Crystal structures of fats and fatty acids, in: Crystallization and polymorphism of fats and fatty Acids, Garti N, Sato K (eds), Marcel Dekker Inc, New York p 97
6. Nyberg SC, Luth H (1972) Acta Cryst B28: 2992
7. Nyberg SC, Porworowski JA (1973) Acta Cryst B29: 347
8. Goto M, Asada E (1978) Bull Chem Soc Japan 51: 70
9. Lomer TR (1963) Acta Cryst 16: 984
10. Goto M, Asada E (1980) Bull Chem Soc Japan 53: 2111
11. Teare HMM (1959) Acta Cryst 12: 294
12. Shearer HMM, Vand V (1956) Acta Cryst 9: 379
13. Pieszek W, Strobl CR, Malzahn K (1974) Acta Crystallogr B30: 1278
14. Goto M, Asada E (1978) Bull Chem Soc Japan 51: 2456
15. Malta V, Cellotti G, Zannetti R, Martelli AF (1971) J Chem Soc (B), 548
16. Goto M, Asada E (1984) Bull Chem Soc Japan 57: 1145
17. Abrahamsson S, Larsson K, von Sydow E (1960) Acta Cryst 13: 770
18. Schaerer AA, Busso CJ, Smith AE, Skinner LB (1955) J Am Chem Soc 77: 2017
19. Kobayashi M, Kobayashi T, Tadokoro H (1984) Mol Cryst Liq Cryst 104: 193

20. von Sydow E (1955) Acta Cryst 8: 845
21. von Sydow E (1956) Acta Chem Scand 10, 1
22. Lutton ES (1967) in Fatty Acids, 2nd Edition, Part 4, Markeley KS (eds), John Wiley & Sons Inc., New York, 2583
23. Vand V, Morley WH, Lomer TR (1951) Acta Cryst 4: 324
24. von Sydow E (1955) Acta Cryst 8: 557
25. Larsson K, von Sydow E (1966) Acta Chem Scand 20: 1203
26. Holland RF, Nielsen JR (1962) J Mol Spectrosc 9: 436
27. Holland RF, Nielsen JR (1963) Acta Cryst 16: 902
28. Kobayashi M (1985) J Mol Struct 16: 193
29. von Sydow E (1955) Acta Cryst 8: 810
30. Boistelle R, Simon B, Pepe G (1976) Acta Cryst B32: 1240
31. Kobayashi M, Sakagami K, Tadokoro H (1983) J Chem Phys 78: 6391
32. Kobayashi M, Kobayashi T, Itoh Y, Sato K (1984) J Chem Phys 80: 2897
33. Aquilano D (1977) J Crystal Growth 37: 215
34. Rinaudo C, Aquilano D, Abbona F (1981) J Crystal Growth 53: 361
35. Tasumi M, Shimanouchi T, Miyazawa T (1962) (1963) J Mol Spectrosc 9: 261; 11: 422
36. Tasumi M, Shimanouchi T (1965) J Chem Phys 43: 1245
37. Tasumi M, Krimm S (1967) J Chem Phys 46: 755
38. Kitagawa T, Miyazawa T (1972) Adv Polym Sci 9: 335
39. Snyder RG, Schachtschreider JH (1963) Spectrochim Acta 19: 85
40. Schachtschreider JH, Snyder PG (1963) Spectrochim Acta 19: 117
41. Koyama Y, Ikeda K (1980) Chem Phys Lipids 26: 149
42. Kobayashi M, Kaneko F, Sato K, Suzuki M (1986) J Phys Chem 90: 6371
43. Hayashi S, Umemura J (1975) J Chem Phys 63: 1732
44. Umemura J (1978) J Chem Phys 68: 42
45. Snyder RG (1979) J Chem Phys 71: 3329
46. Carter VB (1970) J Mol Spectrosc 34: 356
47. Kobayashi M, Tadokoro H, Porter RS (1980) J Chem Phys 73: 3635
48. Kobayashi M, Kobayashi T, Tadokoro H, Cho Y, Kaneko F (1986) Makromol Chem Macromol Symp 5, 1
49. Ewen B, Fisher EW, Piesczek W, Strobl G (1974) J Chem Phys 61: 5265
50. Kobayashi M, Kobayashi T, Itoh Y, Chatani Y, Tadokoro H (1980) J Chem Phys 72: 2024
51. Kobayashi M, Kobayashi T, Itoh Y, Sato K (1986) Bull Mineral 109: 171
52. Beckman W, Boistell R, Sato K (1984) J Chem Eng Data 29: 215
53. Sato K, Suzuki K, Okada M, Garti N (1985) J Crystal Growth 72: 699
54. Sato K, Boistelle R (1983) J Colloid Interface Sci 94: 593
55. Sato K, Boistelle R (1984) J Crystal Growth 66: 441
56. von Sydow E (1955) Arkiv Kemi 9: 231
57. Garti N, Wellner E, Sarig S (1980) Krist Techn 15: 1303
58. Wellner E, Garti N, Sarig S (1983) Cryst Res Techn 16: 1283
59. Bailey AV, Mitcham D, Pittman RA, Sumrell G (1972) J Am Oil Chem Soc 49: 419
60. Sato K (unpublished data)
61. Morishita H, Ishioka T, Kobayashi M, Sato K (1987) J Phys Chem 91: 2273
62. Sato K, Kobayashi M, Morishita H (1988) J Crystal Growth 87: 236
63. Baronnet A (1978) Prog Crystal Growth Charact 1: 151
64. Small DM (1986) Physical chemistry of lipids, in the series: Handbook of Lipid Research, vol 4, Hanahan DJ (ed) Plenum, New York, Chap. 7
65. Garti N, Wellner E, Sarig S (1980) Thermochim Acta 37: 131
66. Kobayashi M (unpublished data)
67. Morishita H, Kobayashi M, Suzuki M, Sato K (1987) In Collected Abstracts of 54th Annual Meeting of Chem Soc Japan, Tokyo 402
68. Inaoka K, Kobayashi, Okada M, Sato K (1988) J Crystal Growth 87: 243
69. Sato K (1979) Japan J Appl Phys 18: 1019
70. Verma AR (1955) Proc Roy Soc A228: 34
71. Amelinckx S (1956) Acta Cryst 9: 16
72. Kaneko F, Kobayashi M, Kitagawa Y, Matsuura Y (1988) Proc. ISA-JOCS Congress, vol. 1: 536
73. Walton G (1969) in Nucleation Zettlemeyer AC (ed), New York, Dekker 225
74. Beckmann W, Boistelle R (1984) J Crystal Growth 67: 271

75. Beckmann W, Rouquerol R, Beckmann H (1983) Thermochim Acta 66: 295
76. Cardew PT, Davey RJ (1982) Proceeding of a Symposium on "Tailoring of Crystal Growth", Manchester, Inst Chem Eng, 1.1
77. Davey RJ, Richards J (1985) J Crystal Growth 71: 597
78. Sato K, Okada M (1972) J Crystal Growth 42: 259
79. Cardew PT, Davey RJ (1985) Proc Roy Soc London A398: 415
80. Sato K (1989) in: Morphology and Growth Units of Crystals, Sunagawa I (ed), Terra Pub Co Tokyo, 513.
81. Murata Y, Motomura K, Matsuura R (1978) Mem Fac Sci Kyushu Univ Ser C 11: 29
82. Garti N, Sato K, Schlichter J, Wellner E (1986) Cryst Res Techn 20: 653
83. Beckmann W, Boistelle R (1985) J Crystal Growth 72: 621
84. Garti N, Sato K (1986) J Am Oil Chem Soc 63: 236
85. Beckmann W (1986) J Crystal Growth 79: 797
86. Beckmann W (1986) J Sci Instrum 19: 444
87. Beckmann W, Kammer S, Meier J, Boistelle R (1986) J Crystal Growth 74: 326
88. Beckmann W, Kammer S, Rauls M (1987) Fat Sci Technol 89: 313
89. Sato K, Boistelle R (unpublished data)
90. Kuhn H (1983) Thin Soild Films 99: 1
91. Mistuya M, Taniguchi Y, Akagi M (1983) J Colloid Interface Sci 92: 29
92. Neuman RD, Fereshtehkhou S, Ovalle F (1984) J Colloid Interface Sci 101: 309
93. Uyeda N, Takenaka H, Aoyama K, Matsumoto M, Fujiyoshi Y (1987) Nature 327: 319
94. Mastuzaki F, Inaoka K, Okada M, Sato K (1984) J Crystal Growth 69: 231
95. Agarwal VK, Ugasaki Y, Mitsubayashi H (1976) Japan J Appl Phys 15: 2327
96. Ueda Y, Ashida M (1980) J Electron Microsc 29: 38
97. Ueda Y (1986) Bull Chem Soc Japan 59: 3775
98. Ueda Y (1987) Bull Chem Soc Japan 60: 2011
99. Miki T, Inaoka K, Sato K, Okada M (1985) Japan J Appl Phys 24: L672
100. Yase K, Inoue T, Inaoka K, Okada M (1989) J Electron Microsc. 38: 132
101. Inoue T, Yase K, Inaoka K, Okada M (1987) J Crystal Growth 38: 306
102. Davies M, Malpass VE (1961) J Chem Soc 1048
103. Cevc G, Marsh D (1987) in: Phospholipid bilayers, Wiley, New York
104. Garti N, Sato K (eds) (1988) Crystallization and polymorphism of fats and fatty acids, Marcel Dekker, New York, Chaps 4 and 6

X-Ray Topography of Organic Crystals

Helmut Klapper

Institut für Kristallographie der RWTH, Templergraben 55, D-5100 Aachen, Federal Republic of Germany

X-ray topography is a powerful non-destructive method for the direct observation of defects in nearly perfect crystals. Since organic molecular crystals usually contain 'light' atoms only, they exhibit low X-ray absorption and scattering power. Thus 'thick' crystal plates (up to 3 mm for CuKα radiation) can be studied by transmission topography, and defects appear by broad kinematical contrast. The image widths of dislocation lines, however, are large, usually > 20 μm, leading to limited spatial resolution and to the requirement of high-quality crystals.

The transmission techniques of projection topography and section topography are explained and illustrated by selected topographs of perfect and imperfect organic crystals. Typical defects observed in nearly perfect organic crystals are inclusions, growth striations, defective growth-sector boundaries, grown-in dislocations and glide dislocations. The determination of Burgers vectors of dislocations and the characterization of planar defects (growth-sector boundaries) by section topography are discussed and demonstrated. Topographs of typical representatives of organic molecular crystals, such as benzil, benzophenone, salol and 2,3-dimethylnaphthalene crystals grown from solution, from undercooled melt, by the Bridgman method or by the Czochralski technique, are presented.

Present address: Mineralogisches Institut der Universität Bonn, Poppelsdorfer Schloß, W-5300 Bonn 1, FRG

Crystals 13

1 Introduction

During the last decade organic crystals have increasingly attracted the attention of solid-state physicists and chemists. This interest arose from the detection of fascinating solid-state effects in some of these materials, such as photoconduction, superconduction, optical nonlinearity, stimulated light emission (laser), etc. Research and technical applications require crystals of high quality or at least of a known and well-defined defect state. When speaking of "quality", two categories of crystal perfection have to be distinguished:

"Chemical perfection" (purity) and
"Structural perfection".

Frequently, a high 'chemical perfection' (i.e. high purity) or a definite level of well-defined impurities (dopants) are of deciding importance for the occurrence of certain solid-state effects and their applications. A survey of the influence of impurities on physical properties, and of the methods of purification has been given by Karl [1, 2].

"Structural defects" comprise all deviations from the strict periodic (lattice) arrangement of atoms, ions, or molecules. They occur also in highly purified crystals, except for certain types of lattice perturbations like growth striations or precipitations which arise from impurities and, therefore, appear only in impure crystals. Structural defects can lead to inhomogeneities which may be detrimental to the accurate determination of physical properties and to their application. They arise during growth, during cutting and polishing (mechanical damages), and during processing (e.g., in the production of semiconductor devices). The knowledge of the origin of lattice perturbations is an essential prerequisite for the control of defect generation and for the improvement of crystal perfection.

X-ray topography in its various techniques has proved to be one of the most powerful methods for depicting and characterizing individual defects (e.g., dislocations) in large and nearly perfect crystals. It is a non-magnifying method which leaves the crystal undestroyed, provided the material does not suffer from radiation damage. Despite its rather poor spatial resolution, X-ray topography has been successfully applied to numerous crystals of a wide range of chemical compositions and physical properties, such as semiconductors, metals, oxides, and, last but not least, organic materials. This report presents a survey of the LANG topographic imaging technique (Lang [3]) and its application to organic crystals. For additional reading on X-ray topography and its various applications, the textbook by Tanner [4] and the reviews by Lang [5] and Authier [6, 7] are recommended. A survey of X-ray topographic studies of organic crystals is given by Izumi [8].

This review focusses on "organic molecular crystals", i.e., crystals containing only hydrogen, carbon, nitrogen, and, as the heaviest atom, oxygen. Such crystals are composed of molecules which are held together by weak Van-der-Waals or hydrogen bonds. Due to their comparatively low X-ray absorption and scattering power, they exhibit some special features in X-ray topographic defect imaging, which are quantitatively, though not in principle, different from inorganic materials containing heavier atoms.

2 Classification of Structural Defects in Organic Crystals

Organic crystals contain the same types of defects as inorganic materials. In the following, a brief description of these defects is given under the aspects of X-ray topographic visualization and characterization. It is anticipated that X-ray topographic contrast arises mainly from the lattice distortions accompanying the defects. Thus, perturbations with sufficiently extended strain fields can be imaged by X-ray topography. The possible primary defects can be classified in a number of categories:

Point Defects:

Point defects are lattice vacancies, (self-)interstitials, molecules in a wrong orientation, deviations from stoichiometry, or impurities. The strain field of such individual defects is much too small to be detected by X-ray topography (which is non-magnifying). If, however, point defects occur in high, but locally varying concentrations, inhomogeneous strain results and generates X-ray topographic contrast (see below under "growth striations" and "growth-sector boundaries").

Line Defects: Dislocations

In organic crystals, dislocations are the most frequent and important extended defects, similarly as in inorganic crystals. They have high significance for the growth and plastic deformation of crystals. Besides their mostly unwanted negative effects they may also have merits, e.g., by allowing crystal growth at low supersaturations, or for gettering of impurities.

A dislocation is characterized by the direction of its line and by its Burgers vector, which is, as a rule, one of the shortest lattice translations. Dislocation lines may be straight or follow irregular curves. Line segments parallel to the Burgers vector **b** have 'screw' character, segments normal to **b** "edge" character, and segments inclined to **b** are "mixed" with screw and edge components. For detailed information on geometrical and physical properties of dislocations, the reader is referred to standard textbooks (e.g. Read [9]).

The lattice deformation associated with a (straight) dislocation decreases according to $1/r$ with the distance r from the dislocation core. Thus the distortions surrounding a dislocation line are rather extended and give rise to a clear X-ray topographic line image. In most cases the Burgers vector of the dislocation can be determined, so that a complete characterization of dislocations by X-ray topography is possible.

Planar Defects

Planar defects comprise high- and low-angle grain boundaries, growth striations, growth-sector boundaries, twin boundaries, stacking faults, and antiphase boundaries.

High- and Low-Angle Grain Boundaries

These boundaries are always accompanied by strong lattice distortions. Coherent low-angle boundaries consist of a (regular) array of dislocations which can be imaged and resolved only under very favorable conditions (e.g., low dislocation density). Grain boundaries can also be detected by the lattice tilt of the adjacent grains.

Growth Striations (Fig. 1)

These lattice perturbations are common in crystals of low purity and occur frequently in organic crystals grown from solution or from impure melts. They arise from local variations of the impurity content and are created by fluctuations in the growth conditions, such as changes of temperature, cooling rate, or convection of the solution or the melt. The impurities may be:

—contaminants (or "dopants") of the solvent or of the melt,
—solvent molecules incorporated,
—deviations from stoichiometry.

The regions of different impurity content form layers coinciding with the instantaneous (plane or curved) growth front, i.e. normal to the (local) growth direction. The varying concentration causes local changes of the lattice parameters and tilts of lattice planes which give rise to X-ray topographic contrast.

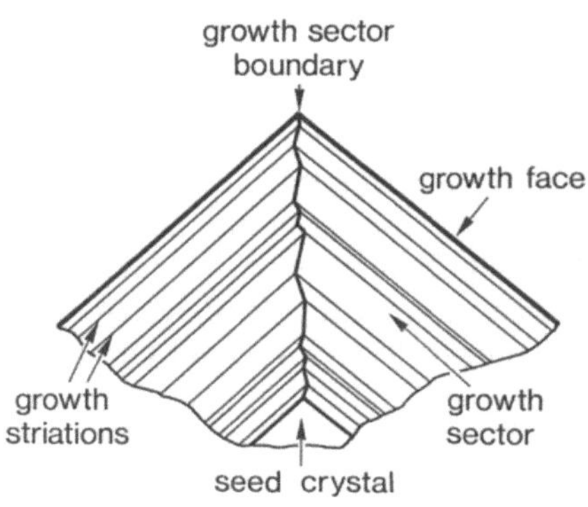

Fig. 1. Growth striations and growth-sector boundary in a crystal grown on planar faces. The growth-sector boundary is the "trajectory" of the edge between the two faces during growth

Growth-Sector Boundaries (Fig. 1)

Growth-sector boundaries separate regions grown on different (planar) faces and thus different growth directions. They are formed by the "trajectory", during growth, of the edge between the two faces involved. Growth-sector boundaries are frequently loaded with impurities and become visible by X-ray topography due to lattice distortions. It is noted that in crystals with pronounced growth striations, the boundaries show an irregular (often zigzag-like) course which indicate fluctuations of the (relative) growth velocities of both faces involved (Fig. 1). The zigzag is correlated to the growth striations. In crystals without (pronounced) striations the growth sector boundary is usually smooth.

Twin Boundaries

Twinning develops during crystal growth, under mechanical load (mechanical twinning) and during phase transitions into lower-symmetry phases. Two categories have to be distiguished: Twins with nonparallel lattices and twins with exactly parallel lattices (merohedral twins). Whereas the first category is rather frequent among organic molecular crystals (in monoclinic crystals they are mostly reflection twins), the latter category is very rare. The well-known Brazil and Dauphiné twins of quartz were not found in benzil $(C_6H_5CO)_2$ ("organic quartz") though it crystallizes in the same enantiomorphic space groups $P3_12$ and $P3_22$.

Twin domains and twin boundaries can be studied and characterized by X-ray topography [10]. Twins with nonparallel lattices, however, are more easily detected by optical methods using polarized light.

Twinning is not treated in the present report. For detailed information on this subject the reader is referred to another review by the author [10].

Stacking-Faults and Antiphase Boundaries

These planar defects, though rather frequent in (structurally simple) inorganic crystals, have not yet been observed in organic molecular crystals. In order to make this survey on crystal defects complete, they are, nevertheless, shortly described:

Stacking faults and antiphase boundaries separate regions (domains) with exactly parallel, but not coinciding lattices, i.e., the lattices are shifted against each other by a "fault" vector. Antiphase domains arise during transitions to phases with lower translational symmetry, e.g., if the centring of the unit cell is suppressed or if a lattice parameter is doubled. Though stacking faults and antiphase boundaries are mostly strain-free, they can be detected with *X*-ray topography by the dynamical fringe contrast ("stacking fault contrast"). The fault vector can be determined.

Volume Defects

Inclusions (bubbles) and precipitates are classified as volume defects. Inclusions arise during growth. In solution-grown crystals they mainly contain mother solution and/or precipitates of impurities. In melt-grown organic crystals, besides precipitates of solute impurities, frequently gas bubbles are formed if growth is not performed under vacuum. This is due to a frequently rather high gas solubility in the melt. The bubbles precipitate when the growth velocity (e.g., during fluctuations of growth conditions) becomes too high. Sometimes bubbles sticking to the growth face pick up more gas from the melt and produce channels when they advance with the growth front [11].

Inclusions are mostly surrounded by severe lattice distortions which generate strong X-ray topographic contrast, provided the inclusions are not too small ($> 1\ \mu m$).

Mechanical Defects (Damages)

Besides the above defects, classified according to their extension, damages due to improper handling and storing of the crystals also should be mentioned: cracks, scratches, and indentations. Cracks may arise from maltreatment or from inhomogeneous temperature changes. These "mechanical defects" appear by high X-ray topographic contrast and obscure other defects of interest, e.g., dislocations. Thus they have to be strictly avoided by proper preparation and handling (cf. Sect. 10).

3 Projection Topography

3.1 Principle

All X-ray topographic methods are based on the reflection of X-rays by a set of lattice planes (*hkl*) of a crystal. The relation between the reflection angle (Bragg angle) θ, the interplanar spacing d_{hkl} and the wavelength λ is given by the Bragg equation

$$2d_{hkl} \cdot \sin\theta = \lambda$$

Two reflection geometries can be distinguished. These are shown and explained in Fig. 2. In the *Bragg case*, which is employed for compact crystals, only defects within a restricted depth below the surface are recorded. This depth is determined by the absorption and the extinction of X-rays. In the *Laue case*, however, all defects within the crystal volume are recorded, provided the absorption is small enough to permit sufficient transmission through the crystal. For organic

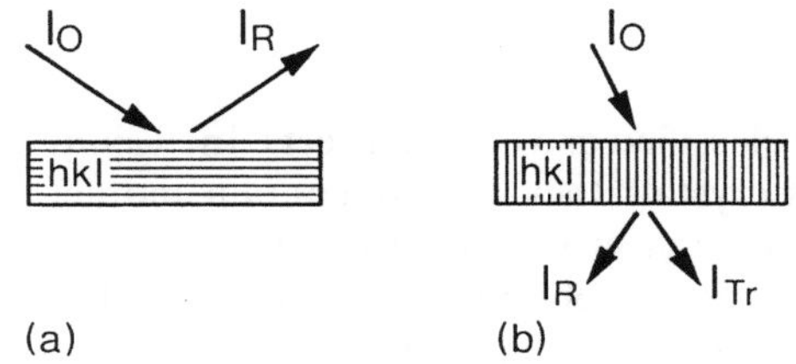

Fig. 2a. Symmetrical Bragg or "reflection" case. **b** Symmetrical Laue or "transmission" case. I_O, I_R, and I_{Tr} are the primary, reflected, and transmitted intensity. The lattice planes may also be inclined with respect to the plate surfaces (asymmetric cases)

crystals with their rather low X-ray absorption (cf. Sect. 3.5) the Laue case is adequate.

The intensity I_R of the reflected beam is mainly determined by the structure factor $F(hkl)$ of the reflection used. The strongest reflections are selected for X-ray topography. Besides the advantage of short exposure times, strong reflections yield narrow defect images (cf. Sect. 6.4) and, consequently, a high spatial resolution of the defects on the topographs.

3.2 The Projection Topographic Method

Among the various X-ray topographic imaging techniques the Lang method [3] is most widely used because it yields the best spatial resolution with conventional X-ray sources (X-ray tubes, rotating anode generators). Its principle is shown in Fig. 3. X-rays from a point focal spot are collimated to a narrow "ribbon-like" (primary) beam which impinges on the crystal plate. The crystal, which is

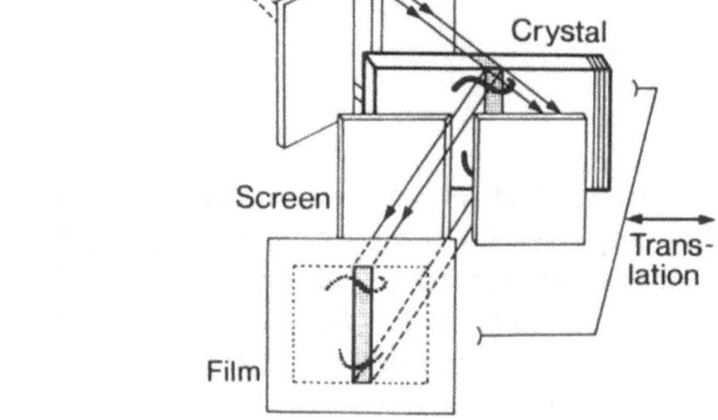

Fig. 3. Schematic view of the Lang technique (projection topography)

preferably (but not necessarily) cut to a plate, is adjusted in such a way that the characteristic Kα_1 line is reflected through the crystal (Laue case) by a selected set of lattice planes ("reflection *hkl*"). The reflected beam is recorded on a film which should be placed as close to the crystal as possible. An adjustable aperture of a screen (Fig. 3) between crystal and film is made so small that only the reflected beam can pass, whereas the transmitted (primary) beam and the incoherent (background) radiation from the crystal are absorbed. The crystal-to-film distance can thus be reduced to a few (typically 10) millimeters.

By simultaneous translation of crystal and film – the incident beam and the aperture of the screen remain fixed – the whole crystal is scanned through the X-ray beam and projected onto the film. Lattice defects are recorded by local changes ("contrasts") of the (integral) reflected intensity. For the low-absorbing organic crystals the kinematical contrast ("extinction contrast") is predominant, which leads to a local increase in reflected intensity.

With modern Lang cameras, crystal plates with diameters up to 12 cm, or in special constructions for 8-inch Si wafers even up to 20 cm can be studied. The (horizontal) primary beam divergence (i.e., in the scattering plane defined by the incident and reflected beam) is kept so small (about 1′) that simultaneous reflection of both lines of the Kα doublet is avoided (see Sect. 3.3), but is still large enough to cover the range of lattice misorientations in the neighborhood of defects (e.g., dislocations), thus leading to the kinematical defect contrast mentioned above. Among the wavelengths available with conventional X-ray tubes, CuKα radiation ($\lambda = 1.54$ Å) is in most cases adequate for the low-absorbing organic crystals.

3.3 Instrumental Image Broadening

A small volume element of the crystal is imaged on the film as a spot of finite diameter. In the Lang method, "image broadening" results from:

- —The finite size of the X-ray focus, in particular of the focus "height" (normal to the scattering plane),
- —the intrinsic angular (perfect-crystal) reflection width w,
- —the spectral width $\Delta\lambda$ of the Kα_1 line,
- —the simultaneous reflection of both lines of the Kα doublet.

The focus height H induces a "vertical" image broadening (normal to the scattering plane):

$$h = (H/L)\cdot D$$

(where L is the distance focus-crystal, D the distance crystal-film). For $H/L = 2 \times 10^{-4}$ (i.e., for a 0.4-mm focus and $L = 2$ m or a 0.1-mm focus and $L = 0.5$ m) and typically $D = 10$ mm, a vertical broadening $h = 2\,\mu$m results.

The reflection width w, the spectral width $\Delta\lambda$ of the $K\alpha_1$ line and the simultaneous reflection of both doublet lines produce a "horizontal" image broadening (in the scattering plane). The reflection width w is very small, $w < 3''$ for organic crystals (see Sect. 6.2), and thus negligible. The broadening resulting from $\Delta\lambda$ is:

$$D\cdot(\Delta\lambda/\lambda)\cdot\tan\theta \approx 1\ \mu m$$

for CuKα radiation ($\Delta\lambda$ is roughly 5×10^{-4} Å), a θ-value smaller than 15° and a crystal-film distance $D = 10$ mm. Simultaneous reflection of the $K\alpha_1$ and $K\alpha_2$ lines, however, generates two images separated by about 7 μm ($D = 10$ mm, $\theta = 15°$, $\Delta\lambda_{1,2} = 3.8 \times 10^{-3}$ Å for the CuKα doublet) and should, therefore, be avoided by adequate reduction of the primary beam divergence. For small angles $\theta < 6°$ this broadening is small ($< 2\ \mu$m), and both doublet lines can be used without noticeable loss of resolution. This is even advantageous because inclusion of the $K\alpha_2$ line increases the reflected intensity by 50%.

Thus the instrumental image broadening of the Lang method is typically 2 μm, leading to a resolution limit of roughly 500 lines/mm (provided only $K\alpha_1$ is used). This broadening, however, is of minor significance compared with the rather broad defect images of X-ray topography. For organic crystals the image widths of dislocations are typically 25–60 μm if the strongest reflections are used in conjunction with CuKα radiation (see Sect. 6.4). Thus, even higher instrumental broadenings (say 5 μm, e.g., due to larger focal spots or the inclusion of the $K\alpha_2$ line) may not be detrimental to a clear presentation of the crystal defects.

These considerations show that the spatial resolution is limited more by the width of the intrinsic defect images than by instrumental constraints. Only crystals with low defect density can be reasonably studied. Assuming a dislocation image width of 20 μm (which is a low value for organic crystals), the dislocation density must be lower than 500 cm^{-1} (i.e., < 500 cm dislocation-line length per cm^2 of film) in order to resolve (equally spaced) individual dislocations.

3.4 Photographic Recording Material

The best photographic materials for recording X-ray topographs are nuclear emulsions (e.g., Ilford L4 plates) which have a grain size far below 1 μm and sufficient sensitivity to X-rays. Emulsions of 12 or 25 μm thickness are adequate for CuKα radiation. Nuclear plates, however, are very expensive and require careful handling and a special developing procedure (for details, see Lang [5], p. 649). Alternatively, fine-grained X-ray films, such as Structurix D4 (AGFA-GEVAERT), can be used. Although they are inferior to nuclear plates in resolution and sensitivity, they yield good X-ray topographic records of organic crystals with sufficient contrast and resolution.

All topographs presented in this review were recorded on Structurix D4 film. The exposure times were about 20 min per 1 cm translation, using a rotating anode generator operated at 0.9 kW with a focal spot of 0.1 mm effective diameter (distance focus-crystal about 50 cm). In order to avoid doubling of the contrasts due to photographic coatings on both film sides, the film was aligned strictly normal to the reflected beam. In a few cases of long exposures only one side of the film was developed.

3.5 Absorption and Specimen Thickness

The comparatively low absorption of organic materials facilitates and accelerates the X-ray topographic recording of crystal defects considerably. For crystals containing H, C, N, and O (as the heaviest atom), the linear absorption coefficients are typically about

$\mu_0 \approx 0.7\,\text{mm}^{-1}$ for CuKα radiation, and

$\mu_0 \approx 0.09\,\text{mm}^{-1}$ for MoKα radiation.

Accepting a reduction of the transmission to $\exp(-\mu_0 t) = 20\%$, crystals with thicknesses t of up to 2.5 mm for CuKα radiation and up to 18 mm for MoKα radiation can be investigated. The use of MoKα radiation, however, is not advisable because of the large widths of the defect images and the low spatial resolution caused by the short wavelength $\lambda = 0.709$ Å (cf. Sect. 6.4). In any case, for specimens with high defect density the thickness must be kept small in order to achieve resolution of individual defects. Thus, CuKα radiation with its narrow defect images is usually the better choice for organic crystals. Moreover, due to a higher photon flux for a given tube voltage and a higher photographic efficiency, exposure times are considerably shorter when using Cu instead of Mo radiation.

4 Section Topography

In projection topography, the "integral reflected intensity", i.e., the reflected intensity integrated over the whole reflection range (cf. Sect. 5.1) is recorded at each point of the film. This implies the superposition of many interference and contrast effects, thus veiling contrast details and suppressing information about the defects involved. This is overcome by the technique of "section topography" [4–7, 12, 13], which is (due to the fine point focus) simply realized in a LANG apparatus by inserting a narrow collimating slit of typically 10 μm width in front of the crystal (Fig. 4). The crystal is not moved during exposure. The X-ray wave-fields excited by the incident beam fill the so-called Borrmann fan (Borrmann

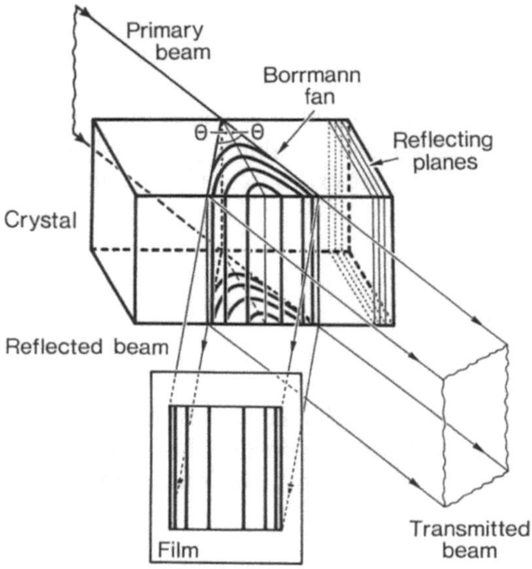

Fig. 4. Schematic view of the section topography technique

triangle) with apex angle 2θ and width $2t \cdot \tan\theta$ on the exit surface of the crystals plate (t: plate thickness). The width of the section topograph on the film (which is placed normal to the reflected beam) is $2t \cdot \sin\theta$. The intensity pattern on the topograph corresponds to the intensity distribution of the interfering reflected wavefield components at the crystal exit surface (see textbooks on: dynamical theory of X-ray diffraction, e.g., James [15], Pinsker [16], cf. Sect. 5.1 of this review).

In the case of a perfect crystal, the reflected-intensity distribution in the Borrmann fan consists of hyperbolic cylinders (surfaces of maximum intensity) as is shown in Fig. 4. Thus, the section topograph of a perfect parallel-sided crystal plate exhibits "vertical" interference fringes (spherical-wave fringes, "Kato fringes") which are crowding closer together towards the margins of the topograph [17, 18].

The advantages of section topography compared with projection topography are the following:

—section topographs reveal the Pendellösung fringes ("Kato fringes") and their distortions due to (slight) lattice strain also for parallel-sided crystal plates (cf. Sect. 5.1),

—they allow the separate recording of the various images associated with defects, i.e., the "kinematical" (or "direct"), the "intermediary" and the "dynamical" image (Authier [6, 12, 13], cf. Fig. 7c, d of this review).

—they exhibit the location of defects in the depth of the specimen plate (cf. Figs. 7c, d).

—they elucidate the character of planar defects, e.g. they distinguish between shift boundaries (stacking faults) and tilt boundaries (cf. Sect. 9).

—the available theory permits computer simulation of section patterns of dislocations and planar defects (stacking faults, twin boundaries) [19].

Section topography, however, has disadvantages too, in that it yields information only from the restricted crystal volume intercepted by the 10 μm primary beam and, to some extent, from the volume of the Borrmann fan. In order to gain such information from a larger part of the specimen, a series of section patterns after stepwise shifts of the crystal on the traverse sled has to be recorded (e.g., on one film, if an overlap of section patterns is avoided). This procedure, however, is rather tedious. Thus section topography will not replace traverse topography but rather supplement it: Section patterns will be recorded only from regions containing such defects, for which more detailed information about their character and location is required.

For more detailed reading about section topography and its merits, the reader is referred to the review by Authier [13].

5 X-Ray Topographs of Perfect Crystals

5.1 Dynamical Interference Phenomena

Before discussing the local intensity changes ("contrast") originating from crystal defects, the topographs of defect-free perfect crystals have to be considered. They are governed by "Pendellösung" phenomena, which always appear on section topographs. This is shown in Fig. 4: The reflected-intensity distribution in the Borrmann fan consists of hyperbolic cylinders (surfaces of maximum intensity). The cut of this distribution with the exit face of the crystal represents the intensity pattern of the section topograph on the film, which – for a parallel-sided plate – exhibits "vertical" interference fringes (spherical-wave fringes, "Kato fringes" [17, 18]) accumulating towards the margins of the topograph. For tapering crystals with wedge-like edges, hyperbolic fringes with the sides of the Borrmann triangle as asymptotes appear. This is demonstrated in Fig. 5.

If crystal and film are translated through the primary beam (i.e., during the exposure of a projection topograph), the intensity of the section topograph is integrated across the width of the pattern, thus recording the local "integrated reflected intensity" on the film. By this integration the fringes are blurred and vanish except for those fringe segments which are parallel to the translation direction, i.e., the segments at the vertex of the hyperbolae in tapering crystal regions (Fig. 5). They lead to Pendellösung fringes of equal crystal thickness ("Ewald fringes"). Thus, the projection topographs of perfect crystals show spatially homogeneous reflected intensities from parallel-sided plates and equal-thickness fringes of increased intensity from tapering regions. Of course, the spacing of the fringes depends on the wedge angle: the fringes are dense for large angles and widely spaced for small angles.

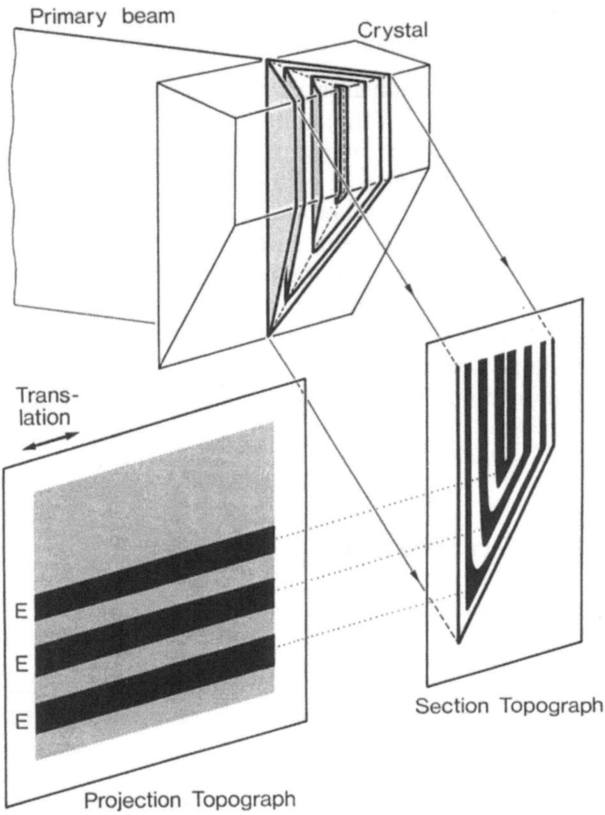

Fig. 5. Distribution of reflected intensity in the Borrmann fan, at the exit surface, on the section topograph, and on the projection topograph of a parallel-sided and a wedge-shaped crystal. Three hyperbolic cylinders (surfaces of maximum reflected intensity) are shown. The section topograph exhibits the typical "spherical-wave" fringes ("Kato fringes"). During translation of the film (projection topograph) an intensity which results from integration across the section topograph is recorded. The section fringe segments parallel to the translation direction produce the equal-thickness Pendellösung fringes *E* ("Ewald fringes") of maximum (relative) intensity.

5.2 Some Theoretical Facts. Extinction Length

Before the presentation of some experimental examples, the essential theoretical features of the Pendellösung phenomena are shortly outlined. The "Pendellösungen" – this term was introduced on 1917 by P. P. Ewald – result from the superposition of two dynamical wave-fields having slightly different wavelengths ($\Delta\lambda/\lambda = 10^{-7} - 10^{-4}$, see textbooks on: dynamical theory of X-ray diffraction, e.g., [15, 16]). The reflected (partial) waves start at the crystal entrance surface with opposite phases. Their interference leads to beats with a period of

$(10^4 - 10^7)\lambda$ (where λ is the X-ray wavelengths), which sums up to some tenths of a millimeter. Thus the beat period ("Pendellösung period") has macroscopic dimensions, and the "Pendellösung" phenomenon is easily detected on X-ray topographs even without any magnification.

For the above beat period the term "extinction distance" is usually applied. In the symmetrical Laue case (reflecting planes normal to the crystal plate) it is the distance between two successive vertices of the maximum-intensity hyperbolae in the Borrmann fan, corresponding (approximately) to the distance between two planes of maximum integral reflected intensity inside the crystal plate. The extinction distance ξ (modified by geometrical factors like wedge angle) controls the spacing of the Pendellösung fringes on projection topographs. According to the dynamical theory of X-ray diffraction it is:

$$\xi \sim (C \cdot \lambda \cdot F/V)^{-1}$$

(for the symmetrical Laue case, disregarding constants and setting the term $\cos\theta \approx 1$, with $C = 1$ for n-polarization, $C = \cos 2\theta$ for p-polarization; F is the structure factor modulus, V the volume of the unit cell).

The extinction length ξ is inversely proportional to the wavelength and the scattering power F/V per unit volume of the reflection used. Representative values of ξ for the strongest reflections of Si are 15–20 μm for CuKα and about 40 μm for MoKα radiation (cf. Lang [5], p. 666, Pinsker [16], p. 54). For organic molecular crystals with only light atoms (usually with oxygen as the heaviest) the scattering power F/V is considerably smaller than for inorganic materials with heavier atoms. Consequently, the extinction distances are larger. Some typical values of ξ for the strongest reflections of benzil, $(C_6H_5CO)_2$, are presented in column 4 of Table 1 (Sect. 6.2). They are roughly 80 μm for CuKα and 200 μm for MoK radiation. Thus, Pendellösung fringes appearing on X-ray topographs of organic crystals are widely spaced and easily detected even under conditions of low instrumental resolution. Thus, in connection with their frequently high perfection, organic crystals are very suitable for the study and demonstration of dynamical X-ray interference phenomena.

5.3 Examples. Polarization Effects

Figure 6 presents two projection topographs of a slightly tapering perfect (0001) plate of benzil recorded in the reflection $20\bar{2}0$ with CuKα and MoKα radiation. The larger spacing of equal-thickness Pendellösung fringes for the shorter Mo wavelength is apparent. Corresponding section topographs, revealing the hyperbolic Kato fringes, are shown in Fig. 7a, b. Note that the width of the section topograph recorded with MoKα radiation ($\theta = 5.6°$) is smaller than that recorded with CuKα radiation ($\theta = 12.26°$), which is due to the smaller Bragg angle θ for MoKα radiation.

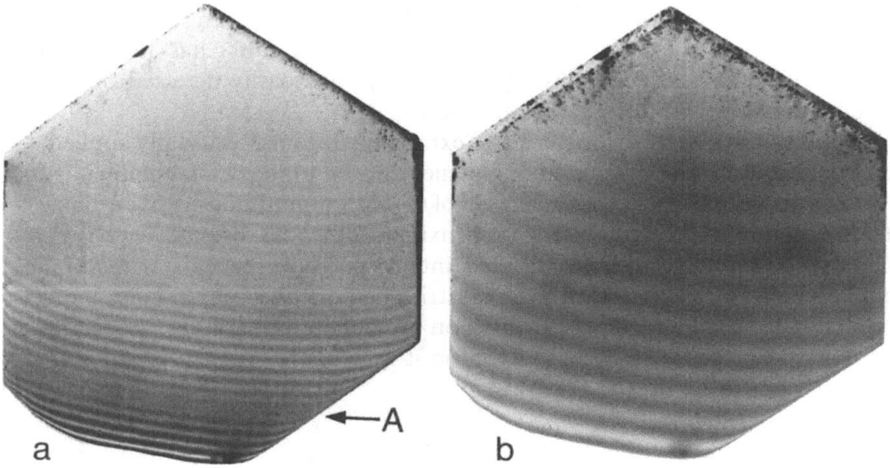

Fig. 6 a, b. Projection topographs of a slightly tapering (0001) plate of trigonal benzil, $(C_6H_5CO)_2$, grown from undercooled melt. The horizontal diameter is about 30 mm. The thickness increases from zero at the bottom to about 2.5 mm at the top edge of the plate. The crystal is defect-free except for some mechanical damages (scratches) near the upper edges. Both topographs were taken in the reflection $20\bar{2}0$. **a** CuKα radiation. **b** MoKα radiation. Note the fading of the Pendellösung fringes in (**a**) after about 5 fringes from the (zero-thickness) bottom edge (see *arrow A*). Due to the higher absorption of CuKα radiation in the thicker parts of the crystal, the upper half in (**a**) shows lower intensity

It is noticed that, on the projection and the section topographs using CuKα radiation, the interference fringes become invisible at certain specimen thicknesses. This results from the fact that X-rays are transversal waves implying two independent polarization states, normal and parallel to the dispersion (incidence) plane (n-polarization and p-polarization). Both states have different Pendellösung periods (extinction distances), the period for p-polarization being smaller by the factor $\cos 2\theta$. The superposition of both Pendellösung oscillations, which start with equal "phase" at zero specimen thickness, reveal another "beat" effect by smoothing out the fringes due to the coincidence of the intensity maxima of the p-oscillations with the intensity minima of the n-oscillations and vice versa. According to theory, the number N of intensity maxima between such "fadings" is [20, 21]:

$$N = \tfrac{1}{2}(1 + \cos 2\theta)/(1 - \cos 2\theta).$$

Thus, the fringes vanish for the first time (counted from zero thickness) after $\tfrac{1}{2}N$ maxima. For CuKα radiation ($\theta = 12.26°$) this is after about 5 fringes which is in good agreement with the observation (Fig. 6a). For MoKα radiation the first

Fig. 7 a, b. Section patterns corresponding to the projection topographs of Fig. 6a, b (but with higher magnification), taken with reflection $20\bar{2}0$ **a** with CuKα radiation, **b** with MoKα radiation. Note in (**a**) the fading of the hyperbolic fringes at two thicknesses (marked by arrows). The lower

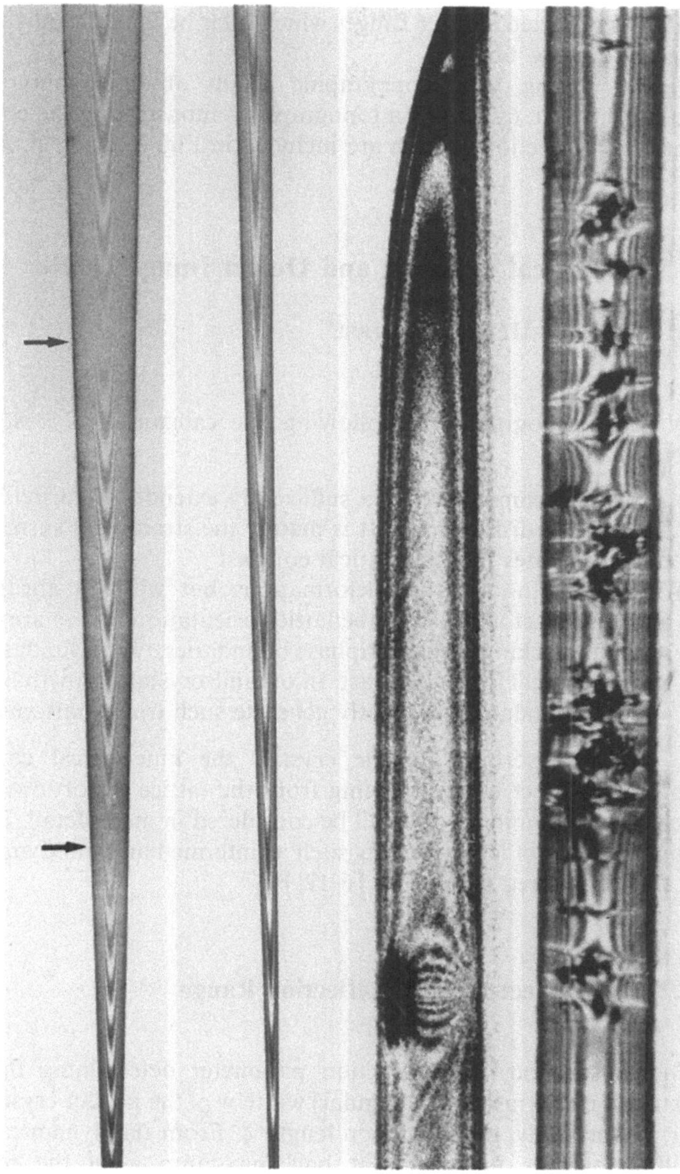

intensity in the upper part of the pattern (**a**) (higher absorption) was balanced by increased exposure during optical reproduction. **c** Section topograph of a parallel-sided benzil ($1\bar{1}00$) plate (1 mm thick) with tapering (rounded) upper edge. Reflection 0003, CuKα. Label D: screw dislocation (cf. Sect. 6.5). **d** Section topograph of a benzil ($11\bar{2}0$) plate, reflection $2\bar{2}00$, CuKα, showing many dislocations (roughly) normal to the section (cf. Sect. 6.5)

fading is expected after 26 fringes which is far beyond the maximum thickness of the crystal (Fig. 6b).

Anticipating X-ray topographic results obtained for crystals containing defects (cf. Sect. 6.5), section topographs demonstrating the effect of dislocation lines on the section pattern are included in Fig. 7c, d for direct comparison.

6 Kinematical Contrast and Defect-Image Width

6.1 Types of Defect Contrast

By X-ray topography the following two categories of crystal defects can be investigated:

i) Defects accompanied by a sufficiently extended deformation of the crystal lattice (e.g. dislocations). It is mainly the strain field surrounding the defect that generates the kinematical contrast.
ii) Defects without lattice deformations but with an abrupt change of the structure factor phase or the lattice orientation. These are essentially planar defects (stacking faults, antiphase boundaries, twin boundaries), which appear by dynamical fringe contrast. In organic crystals, growth bands and growth-sector boundaries frequently generate such fringe patterns.

In low-absorbing *organic crystals* the kinematical contrast (extinction contrast, "direct image") arising from the lattice distortions around a crystal defect is predominant and will be considered in more detail. For other contrast contributions of lattice defects such as intermediary and dynamical images, the reader is referred to Authier [6, 12, 13].

6.2 The Perfect-Crystal Reflection Range

The most important diffraction parameter determining the kinematical or extinction contrast is the (angular) width w of the perfect-crystal reflection curve or, alternatively, the extinction length ξ. From the dynamical theory of X-ray diffraction, the full width at half maximum w of the reflection curve is (disregarding constants and setting the term $\cos\theta \approx 1$):

$$w \sim \lambda \cdot d \cdot F/V \quad \text{and} \quad \xi = 2d/w$$

where d is the interplanar spacing, F the structure factor modulus, V the volume of the unit cell. For organic crystals with only light atoms (usually with oxygen as the heaviest), the scattering power $d \cdot F/V$ per unit area of the reflecting lattice

Table 1. Angular reflection ranges w (full width at half maximum) and extinction lengths ξ for three selected reflections of benzil, using CuKα and (in brackets) MoKα radiation (symmetrical Laue case, π-polarisation)

hkil	F	$w['']$	$\xi[\mu m]$
$2\bar{2}0\bar{1}$	82.8	2.04 (0.90)	71 (162)
$2\bar{2}00$	61.9	1.57 (0.71)	96 (210)
0003	21.1	0.67 (0.30)	280 (628)

plane is considerably smaller than for inorganic materials containing heavier atoms. Thus, rather small values of w result even for the strongest reflections and CuKα radiation. This is the reason for the broad kinematical contrasts ("direct images"), as will be explained below (Sect. 6.4). In Table 1 the values of F, w, and ξ are listed for three reflections of trigonal benzil. For benzil, $(C_6H_5CO)_2$, the reflection $2\bar{2}0\bar{1}$ has the largest structure factor and the widest reflection range w for a given wavelength. It is noted that the reflection ranges for MoKα radiation are only about half of those for CuKα radiation.

6.3 The Origin of the Kinematical Contrast

The perfect-crystal reflection widths w of the strongest reflections are of the order $1''$ for CuKα radiation. The incident beam divergence of the Lang technique, however, is $1-3'$. This means that only a very small angular sector of the incident beam is reflected by the perfect crystal. The rays outside this sector propagate through the perfect regions "off resonance". If such "outside" rays meet a distorted crystal region with misoriented lattice planes, for some of them the reflection condition will be locally satisfied, and they will be reflected. Such rays diffracted from distorted regions also propagate through the surrounding perfect regions without resonance, and thus do not suffer (primary) extinction ("extinction contrast").

In the distorted crystal regions the misorientations range continuously from zero (at the periphery of the strain field) up to some minutes of arc (at the center of the defect). Thus, the rays reflected in distorted regions cover an angular sector of the incident beam which is much larger than the narrow w-sector of rays reflected in the perfect crystal. This wide sector of incident rays, however, contains much more X-ray intensity which is converted into the reflected beam by the strain field, leading to a considerable increase of diffracted intensity. Thus, defects accompanied by lattice strain appear on X-ray topographs by locally increased intensity. It is emphasized that the prerequisite of such kinematical defect contrast is a sufficient divergence of the incident beam.

6.4 The Kinematical Dislocation Image Width

The above considerations yield the key for an estimate of the kinematical image width of defects such as dislocations. In this context the "effective misorientation" $\delta\theta$ of the reflecting lattice planes has to be considered. It is essentially the orientation change of the reflecting planes between the distorted and the perfect region, measured as an angular deviation in the scattering plane (for details, see Authier [6, 12, 13]). This effective misorientation varies from zero at the periphery of the strain field to higher values near the defect center.

Assuming the full width of the perfect-crystal reflection range to be $2w$ (i.e., extending from $-w$ to $+w$ around the exact Bragg angle), the borderline of the kinematical contrast is roughly determined as follows: Rays of the incident beam which are reflected in (peripheral) strain regions with effective misorientation $\delta\theta < w$ propagate within the perfect-crystal reflection range and suffer primary extinction (no kinematical contrast). Rays which are reflected in regions with $\delta\theta > w$ propagate outside the reflection range and generate contrast. Thus the borderline of contrast is governed by the condition:

$$\delta\theta \approx w.$$

Though the transition from the extinction case to the extinction-free case is continuous and not sharply defined, this relation is very helpful for the estimation of defect image widths. It will now be applied to estimate the contrast widths of pure-screw dislocations [22–24] for which the effective misorientation is simply the slope of the helically deformed lattice planes (Fig. 8) at a distance r from the dislocation core (b_g = Burgers vector component parallel to the diffraction

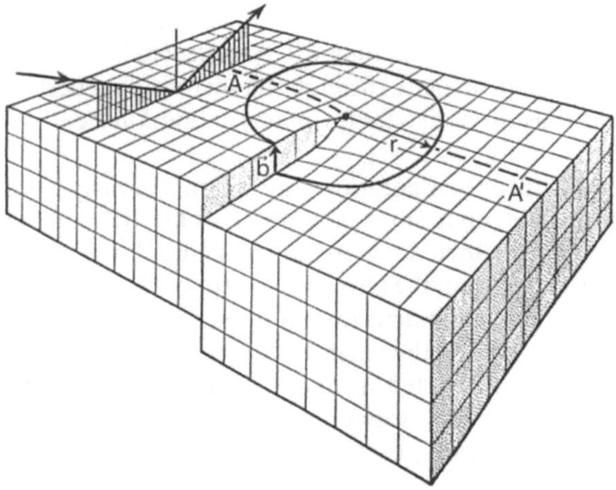

Fig. 8. Screw dislocation. The effective misorientation of the helically distorted reflecting lattice planes (along the line A–A′ normal to the incidence plane) in a distance r from the screw dislocation core is given by the slope $b/2\pi r$ of the helix of radius r

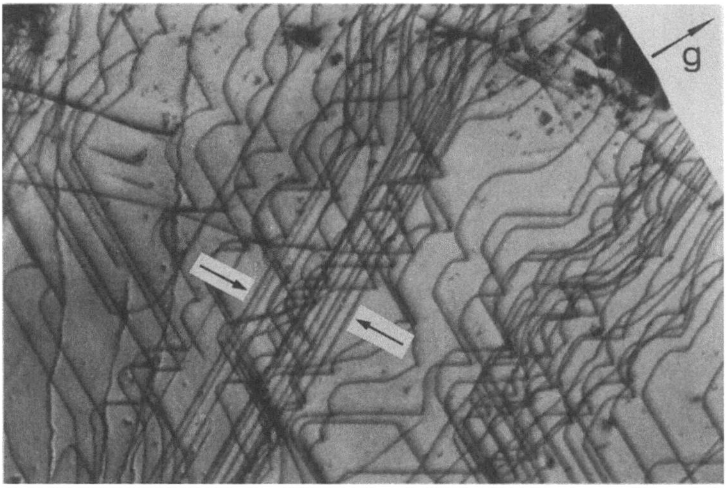

Fig. 9. X-ray topograph (CuKα radiation) of a (0001) plate of benzil grown from an undercooled melt. Thickness: 1.2 mm, section $6.7 \times 9.6\,mm^2$, reflection $2\bar{2}00$ (g: diffraction vector). The straight dislocation lines marked by arrows are grown-in pure-screw dislocations with Burgers vector $b = [100]$ and image widths of about 35 μm. Many other dislocations show post-growth motion impeded by pinning points

vector g):

$$\delta\theta = b_g/2\pi r$$

From this and the above condition the dislocation image width is:

$$V = 2r \approx b_g/\pi w = \xi(g \cdot b)/\pi$$

In Table 2 the calculated image widths of pure screw dislocations in benzil are presented for the reflections of Table 1 for CuKα and MoKα radiation. In three cases the observed widths, as determined from Figs. 9 and 10, are included. The agreement is excellent.

Table 2. Calculated and (in brackets) observed image widths V of pure screw dislocations in benzil with Burgers vectors $\mathbf{b} = [100]$ (8.376 Å) and $\mathbf{b} = [001]$ (13.70 Å) for three reflections with different structure factors, for CuKα and MoKα radiation

hkil	**b**	b_g[Å]	V_{Cu}[μm]	V_{Mo}[μm]
$2\bar{2}0\bar{1}$	[100]	7.02	23	52
$2\bar{2}00$	[100]	7.25	31 (35)	69
0003	[001]	13.7	134 (137)	300 (310)

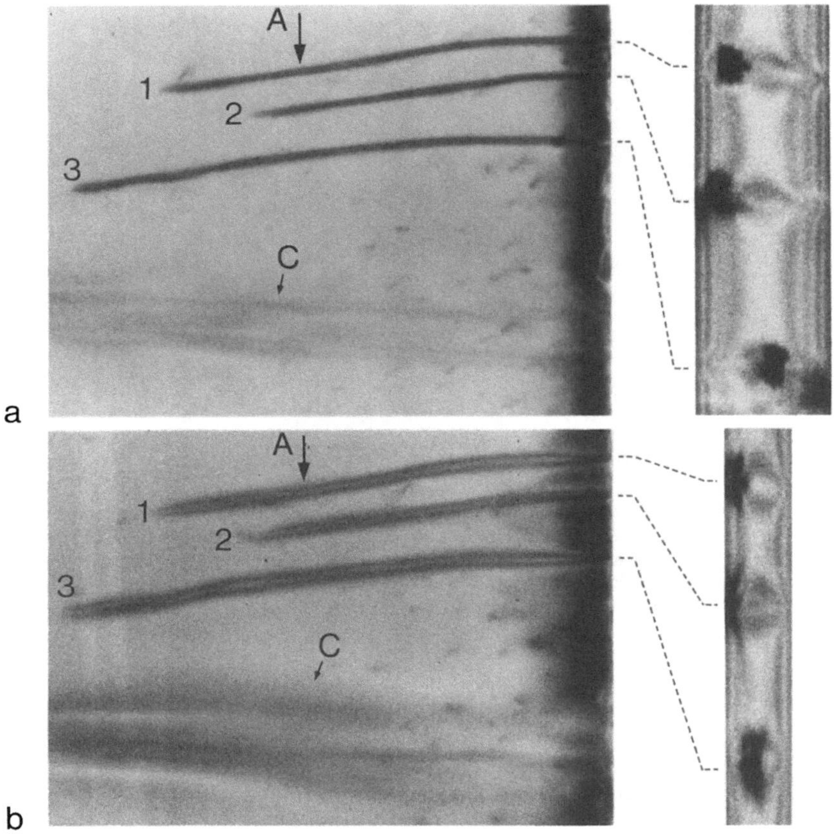

Fig. 10 a, b. Three pure-screw dislocations with Burgers vector [001] along the trigonal axis of benzil (grown from an undercooled melt). At left the lines emerge out of the $(10\bar{1}0)$ plate which has a thickness of about 1.1 mm. Size of imaged field: $4.8 \times 6.3\ mm^2$. Both topographs are taken in reflection 0003. **a** $CuK\alpha$ radiation; the image widths of dislocation lines are 137 μm (average). **b** $MoK\alpha$ radiation; the image widths are 310 μm (average). For the faint contrast C, see text. For the section topographs at right (recorded at the locations marked by the arrow A) refer to Sect. 6.5

It is apparent that the rather broad images result from the extremely narrow reflection ranges which are typical for organic crystals with low scattering density. For the screw dislocation along [001], the long Burgers vector (13.7 Å) contributes considerably to the width of the image. It is also clear that the very wide images for Mo radiation prohibit the use of this wavelength for X-ray topographic studies of organic crystals, despite the very low absorption.

Finally, the attention is directed to the faint contrast C in Figs 10a and b. It arises from a bundle of grown-in pure-edge dislocations along the trigonal axis [001] (Burgers vector ⟨100⟩). Their images are "extincted" according the $g \cdot b$ criterion (cf. Sect. 7.2). The faint residual contrast will be discussed in Sect. 7.2.

6.5 Section Topography of Dislocations

Following the outlines of Sect. 4, some typical features of section topographs with dislocations will be shown here. For more details reference is made to Authier [6, 12,13].

On the right side of Figs. 10a, b the section topographs cutting the three dislocation lines 1,2,3 at the position indicated by the arrow are presented. The corresponding direct (kinematical) contrasts on the projection and on the section topographs are connected by dotted lines. From the position of the direct image within the width of the section topograph it follows that the long dislocation line 3 lies roughly in the center of the plate thickness, whereas the two other lines (in particular the short line 2) are located close to the X-ray entrance surface. Since the exposure time of the section topographs was chosen to yield an adequate record of the fringe pattern, the much more intense direct images of the dislocations are highly overexposed, so that their fine structure (bimodal contrast, see following Sect. 6.6) is obscured. The wider section width obtained with CuKα radiation is due to the larger θ-angle.

The variety of contrasts arising from dislocation strain fields is apparent in Figs. 7c, d. Label D in Fig. 7c marks the contrast of a single pure-screw dislocation normal to the section and lying close to the X-ray entrance surface. Note the distortions of the Kato fringes on the right side of the "direct" dislocation image. Its bimodal contrast (see following Section) is clearly recognized, despite overexposure. The fringe distortions result from those faint lattice deformations (at greater distances from the dislocation core) which are not big enough to generate kinematical contrast.

In Fig. 7d many dislocations running more or less normal through the section are shown. Note the white "dynamical" images (cf. Authier [6, 12, 13]) attached to the "direct" images (dark spots) and the characteristic distortions of the Kato fringes. Many of these various contrast features are obscured by integrating the intensity across the section pattern (i.e., by projection topography).

6.6 The Bimodal Contrast Profile of Dislocation Images

The X-ray topographic kinematical images of dislocations sometimes exhibit a bimodal intensity profile [6, 25, 26]. This is a general phenomenon occurring for all kinds of crystals. The bimodal contrast, however, can only be resolved and observed in cases of wide dislocation images and under optimum resolution conditions. For dislocations in organic crystals with their wide images, this double contrast very frequently appears, even under conditions of poorer instrumental resolution, in particular if reflections of medium- or low- structure factor moduli are applied. Examples are given in Fig. 10 (pure-screw dislocations) and Fig. 12 (mixed and pure-edge dislocation).

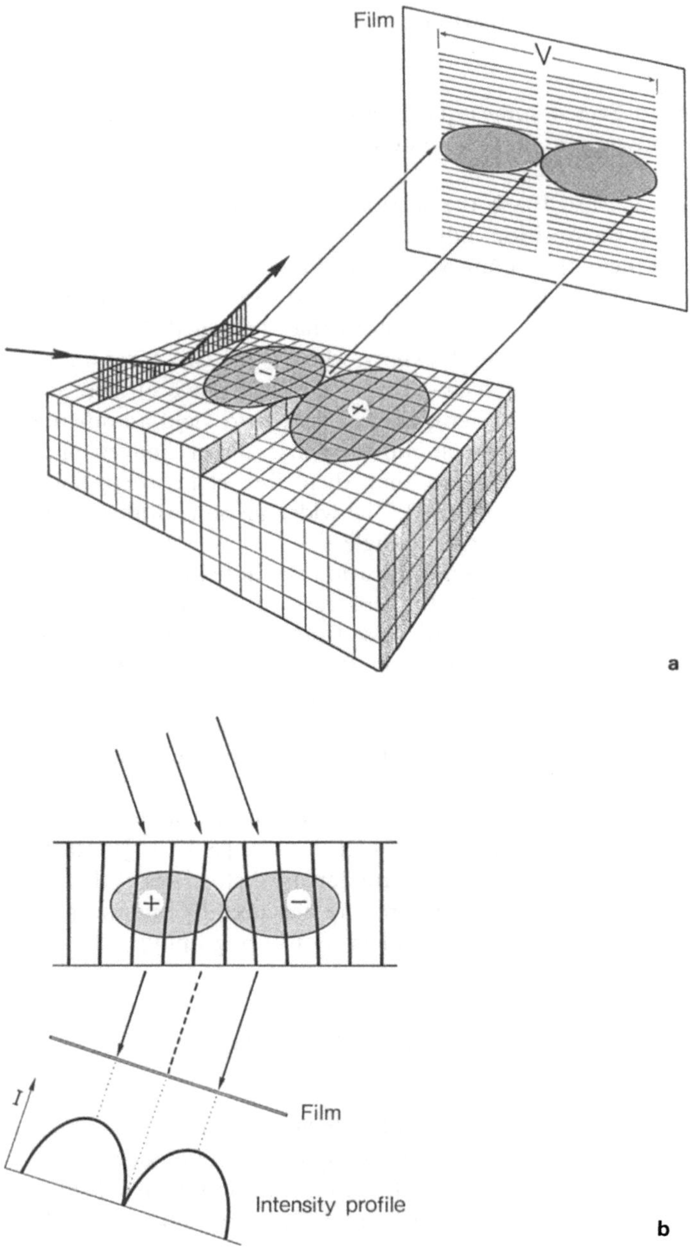

Fig. 11 a, b. Origin of the double contrast of dislocation lines. **a** Screw dislocation, **b** edge dislocation. The *shaded areas* ("columns" along the dislocation lines) are regions in which the effective misorientation of reflecting planes exceeds the angular reflection width *w*, thus leading to kinematical contrast (cf. Sect. 6.6). V: image width of screw dislocation (cf. Sect. 6.4)

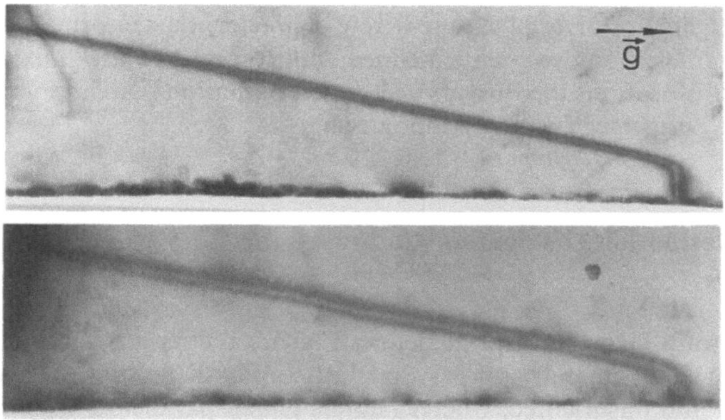

a

b

Fig. 12 a, b. A single grown-in dislocation line in benzil, Burgers vector [001], revealing bimodal contrast, i.e. an intensity minimum in the line center; reflection 0003, depicted crystal area 2.3 × 8.7 mm. **a** CuKα radiation, **b** MoKα radiation. The long inclined line segment has "mixed" (but predominant screw) character, the short vertical segment is pure-edge. At the sharp bend the dislocation line penetrates a growth-sector boundary

The kinematical double contrast is explained as follows (Fig. 11): At both sides of the dislocation line core (viewed in the X-ray projection direction) two "columns" of strained material exist with the effective misorientation of the reflecting planes exceeding the critical value w (cf. Sect. 6.4). These two "columns" which touch along the dislocation core, have opposite signs of effective misorientation. The kinematical contrast originates from these regions. The intensity projected on the film is roughly given by the primary-beam path-length through the columns. Therefore, the intensity diffracted from the center of the dislocation strain field is small or zero, which also may be due to the very high effective misorientation exceeding the divergence of the primary beam. This results in the intensity dip in the center of the kinematical dislocation image.

7 Determination of Dislocation Burgers Vectors

7.1 General Remarks

Dislocations are formed during crystal growth, by plastic processes (dislocation glide), or by the condensation of vacancies or self-interstitials. With respect to their origin, geometrical features, and physical behaviour, dislocations in organic crystals are quite similar to those in inorganic materials. Review papers on dislocations investigated by X-ray topography are given by Lang [24] and Klapper [27].

Dislocations are line defects which may be straight or irregularly curved. They are characterized by their paths and by their Burgers vectors. Line segments parallel to the Burgers vector **b** have pure-screw character, those normal to **b** have pure-edge character, and segments inclined with respect to **b** are "mixed". In most studies of dislocations the Burgers vector determination is an important task which can be performed by X-ray topography.

7.2 The g·b Criterion for Dislocation Contrast

In a dislocation strain field the displacements of lattice points are parallel to the Burgers vector **b**. From this follows that lattice planes normal to **b** are strongly disturbed and misoriented, whereas lattice planes parallel to **b** remain undeformed. Consequently, a dislocation imaged with reflecting planes normal to **b** (diffraction vector **g** parallel **b**) generates a strong and broad contrast, whereas for reflecting planes parallel to **b** no contrast will appear. For imaging planes inclined to the Burgers vector, the contrast lies between these two extremes.

These considerations are collected in the **g·b** criterion [3, 28], which expresses the fact that the dislocation contrast is proportional to the scalar product $\mathbf{g}\cdot\mathbf{b} = b_g/d$ (b_g: component of **b** parallel to **g**, d: interplanar spacing). Thus, the Burgers vector can be determined by analysis of the dislocation contrast on several topographs taken in different reflections. In particular, if two reflections can be found where the dislocation images are "extinct" ($\mathbf{g}\cdot\mathbf{b} = 0$), the Burgers vector must be parallel to the intersection of the two corresponding planes. Since Burgers vectors are generally the shortest lattice translations, the shortest vector along the direction obtained will be the correct one.

7.3 Examples and Restrictions

An example of the "extinction" of dislocation images due to the $\mathbf{g}\cdot\mathbf{b} = 0$ criterion and of a Burgers vector determination for benzil (grown from undercooled melt) is presented in Fig. 13.

According to the theory, the complete "invisibility" of dislocation images for $\mathbf{g}\cdot\mathbf{b} = 0$ is strictly correct only in special cases, namely for pure-screw and for pure-edge dislocations along any direction in elastically isotropic media and along even-fold symmetry axes in anisotropic materials. (For the very faint contrast of edge dislocations for **g** normal to the glide plane, see Lang [24]). Due to elastic coupling, however, screw and edge dislocations along non-symmetry directions or along a threefold axis, and mixed dislocations along any direction frequently generate contrast despite $\mathbf{g}\cdot\mathbf{b} = 0$. The residual dislocation contrast C in the topographs of Fig. 10 (edge dislocations along a threefold axis) and Fig. 13b (mixed character) are examples.

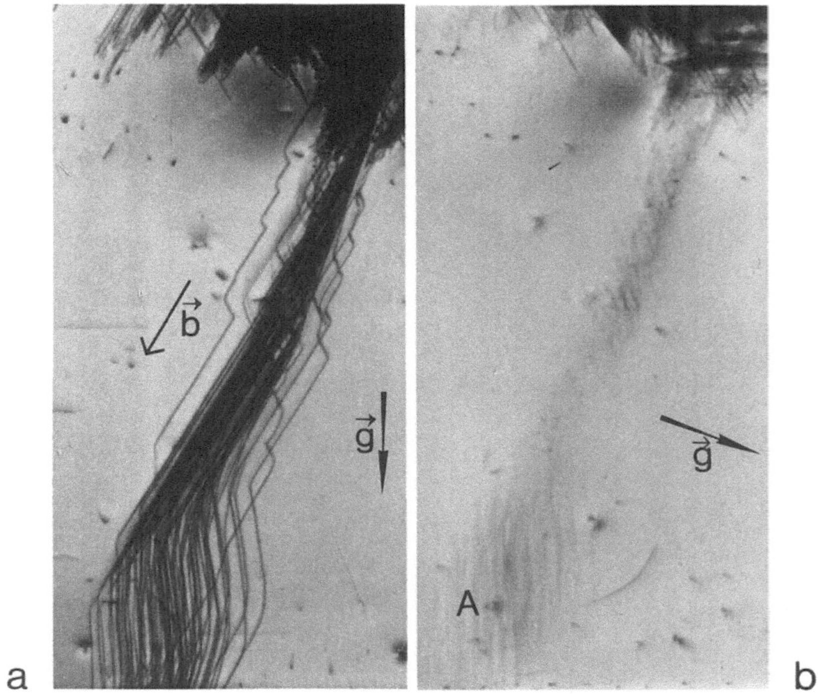

Fig. 13 a, b. Bundle of grown-in (neearly) pure-screw dislocations in benzil with Burgers vector [100], originating from the seed crystal in the upper right corner. **a** Reflection $20\bar{2}0$ with $\mathbf{g}\cdot\mathbf{b} = 2$ (strong contrast), **b** Reflection 0220 with $\mathbf{g}\cdot\mathbf{b} = 0$ (dislocations invisible). Note the residual contrast, in particular at A ("mixed" character)

In many cases the dislocation contrast is not "residual" but rather strong, despite $\mathbf{g}\cdot\mathbf{b} = 0$. This is illustrated by the topographs Fig. 14a, b of salol (orthorhombic) grown from undercooled melt: The plate contains the growth sectors (100) (which grew out), {010} and {120}. There are many dislocations originating from the seed crystal in the (100) growth sector. These penetrate two growth sector boundaries, thus changing their directions (cf. Sect. 8.1.1 two times from [100] (vertical) to [010] (horizontal). Their Burgers vectors are [100], [010], and [001]. Irrespective of the Burgers vectors involved, it is apparent that dislocation segments parallel to the twofold axes [100] or [010] are completely "extinct" according to $\mathbf{g}\cdot\mathbf{b} = 0$ either in reflection 200 or in 020, whereas segments along non-symmetry directions exhibit relatively strong contrast. In these cases, however, the contrast of such segments is usually a minimum compared with reflections which fulfill $\mathbf{g}\cdot\mathbf{b} \neq 0$.

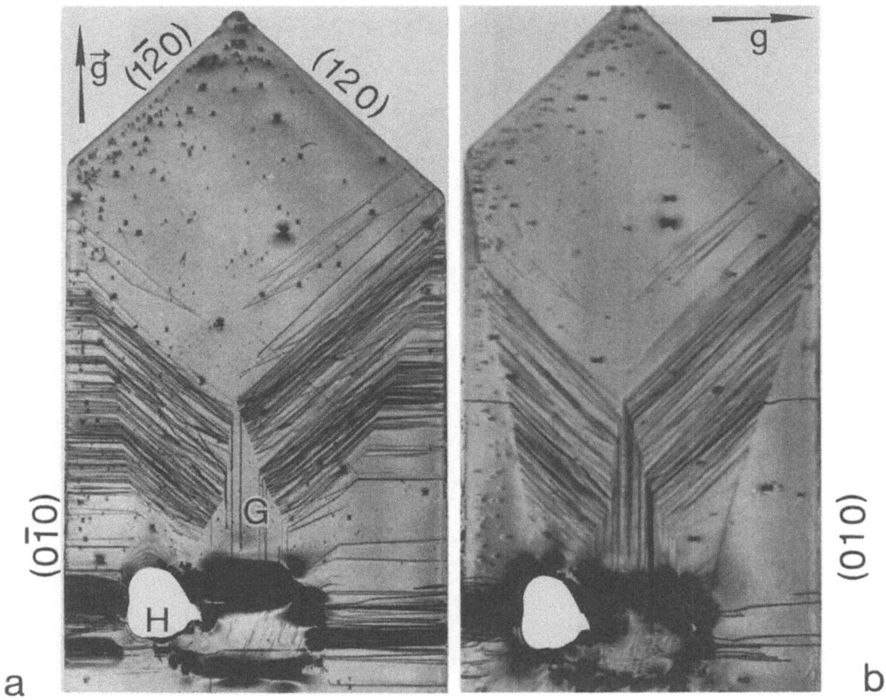

Fig. 14 a, b. (001) Plate of salol ($C_{13}H_{10}O_3$), 1.5 mm-thick, horizontal dimension 26 mm. CuKα radiation. **a** Reflection 200, **b** reflection 020. H: Hole caused by a large gas inclusion (bubble), G: Grown-out (100) growth sector. The dots arise from surface damages (indentations due to dust particles, cf. Sect. 10)

8 Dislocation Configurations

According to their origin, two categories of dislocations have to be distinguished:

—dislocations nucleated during growth, i.e., at the growth face ("grown-in" dislocations),
—dislocations generated after growth, i.e., behind the growth front ("post-growth" dislocations).

The grown-in dislocations may have preserved their "as-grown" geometry and distribution or suffered post-growth movement, e.g., by stress arising during the cooling period after finishing growth, or by dislocation climb. Post-growth dislocations, which may arise in regions "behind" the growth front already during growth, result from plastic deformation under stress, or by the condensation of vacancies or self-interstitials. All these kinds of dislocations have been observed in organic molecular crystals. As will be shown later, "grown-in" and "post-growth" dislocations can be distinguished according to their typical geometry.

8.1 Crystals Grown on Planar Faces

8.1.1 General Features of Grown-in Dislocations

Crystals grow on planar (habit) faces from solution, from undercooled melt and from the vapour phase. As has been mentioned in Sect. 2 (Fig. 1) and is shown in Fig. 15, such crystals consist of growth sectors, i.e., regions grown in different directions, separated by growth-sector boundaries. The configuration of grown-in defects, in particular of dislocations, is strongly influenced by these divisions.

Grown-in dislocations originate from inclusions which frequently occur at the surface of the seed crystals or are generated by growth accidents. As a rule, the dislocations follow straight lines with rather sharply defined directions. These preferred directions (cf. Sect. 8.1.2) which in most cases lie within an angle of $\pm 20°$ to the growth normal **n**, depend on the growth direction and on the Burgers vector involved. The dependence on the growth direction becomes strikingly apparent when the dislocation line penetrates a growth-sector boundary, i.e., when the dislocation end point shifts during growth from one face to the other. According to the abrupt change of growth direction, the dislocation line undergoes a sharp bend and proceeds along a new direction which is characteristic of the new growth sector.

This typical arrangement of grown-in dislocations, which is schematically shown in Fig. 16, has been observed in numerous crystals of various chemical

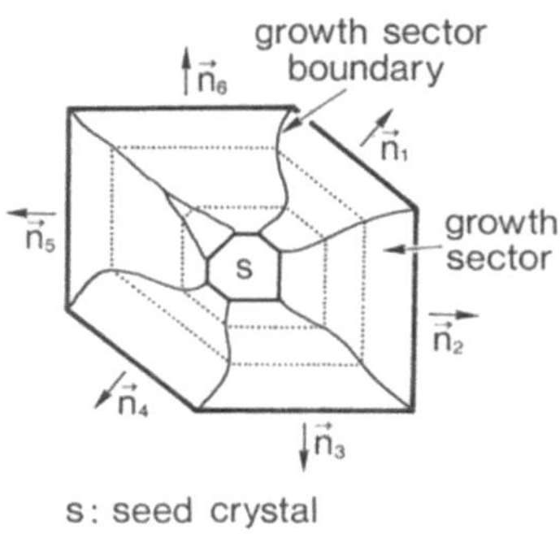

Fig. 15. Division of a crystal into growth sectors. The vectors n_i indicate the growth directions. *Dashed lines*: contours of the crystal at different stages of growth. One of the growth sectors has "grown out"

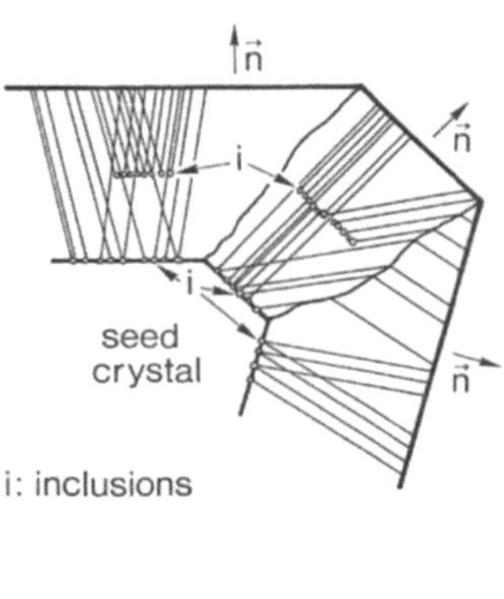

Fig. 16. Typical geometry of grown-in dislocations. The different preferred directions of dislocation lines within one growth sector result from different Burgers vectors

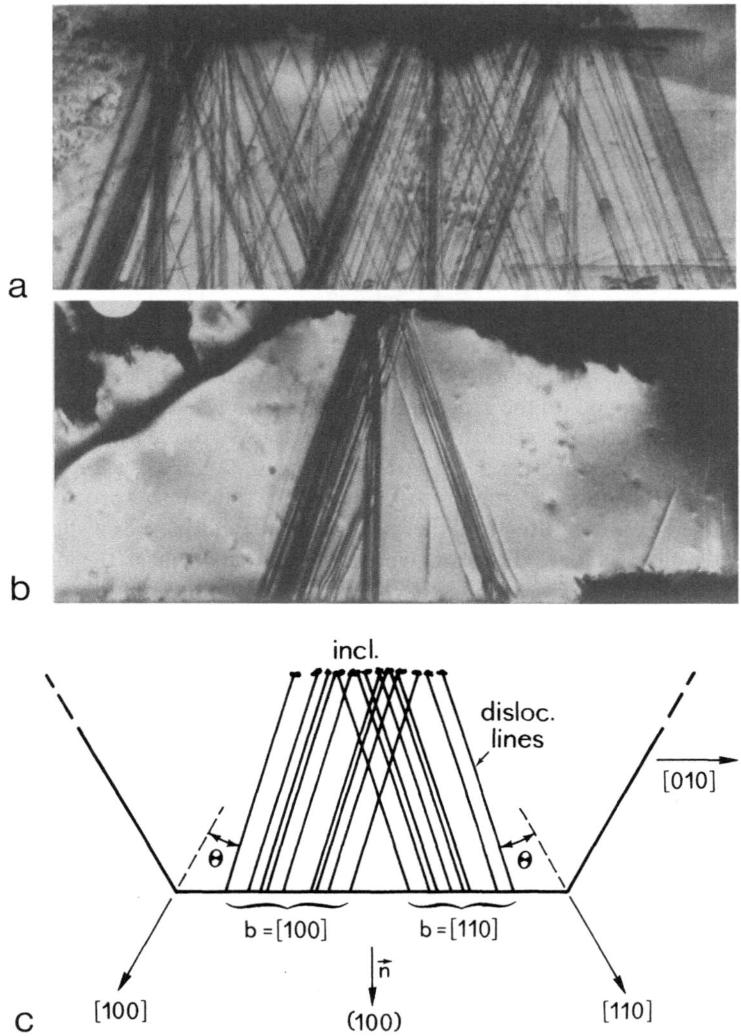

Fig. 17 a–c. Grown-in dislocations in the (10$\bar{1}$0) growth sector of benzil **a** Solution-grown crystal; **b** melt-grown crystal; **c** illustration of dislocation geometry and Burgers vectors (after Ref. [11])

compositions and physical properties, e.g., KDP [29], thiourea [30, 31], ammonium hydrogen oxalate hemihydrate [32] – for an overview, see Refs. [27, 33] – and also in organic molecular crystals (benzil [34, 35], benzophenone [33], salol [27], pentaerythritol tetranitrate (PETN) [36–38], cyclotrimethylene trinitramine (RDX) [38–40]). It is interesting to learn that the preferred directions of grown-in dislocations are independent of the growth method, provided that the same growth sectors (growth directions) are present.

This is demonstrated in the topographs of Fig. 17 (after Scheffen-Lauenroth et al. [11]) showing dislocations with Burgers vectors [100], [010], and [110] in a $(10\bar{1}0)$ growth sector of benzil grown from a solution in xylene (a) at about 40°C and from an undercooled melt (b) at $T_m = 96$°C. In both crystals the three preferred directions are the same. An example of the bending of dislocation lines when passing through growth sector boundaries is found in Fig. 14.

8.1.2 Theory of Preferred Directions

A dislocation line emerging from a growth face proceeds into the newly grown layer and increases the Gibbs' free energy of this layer by the amount of its own energy. The tendency to minimize the Gibbs' free energy requires that the energy of the dislocation within the layer is a minimum. Thus, a straight dislocation line is favored because any bending increases both its energy per unit length and its whole length within the layer. Referring to Fig. 18, the total energy of a straight dislocation line in the layer is:

$$E = E(\mathbf{b}, \mathbf{l}, c_{ij}) \cdot d / \cos \alpha$$

($\mathbf{b}$: Burgers vector, $\mathbf{l}$: unit vector along the dislocation line direction, $\alpha = \alpha(\mathbf{l}, \mathbf{n})$: angle between $\mathbf{l}$ and the growth direction $\mathbf{n}$, d: thickness of the layer). $E(\mathbf{b}, \mathbf{l}, c_{ij})$ is the elastic energy per unit length of the dislocation line which depends also on the elastic constants c_{ij} of the crystal. Here, only the elastic energy is considered because the entropy contribution to the Gibbs' free energy of a dislocation can be neglected. Since the direction of minimum energy does not depend on d, the energy W per unit growth lengths ($d = 1$) is regarded:

$$W(\mathbf{b}, \mathbf{l}, \mathbf{n}, c_{ij}) = E(\mathbf{b}, \mathbf{l}, c_{ij}) / \cos \alpha$$

Now the preferred direction $\mathbf{l}_0$ is given by the condition:

$$W(\mathbf{b}, \mathbf{l}, \mathbf{n}, c_{ij}) = \text{Minimum}$$

It is obvious that $\mathbf{l}_0 = \mathbf{l}_0(\mathbf{b}, \mathbf{n}, c_{ij})$ defined in this way depends on the Burgers vector $\mathbf{b}$ and the growth direction $\mathbf{n}$ and, therefore, accounts for the geometrical features of grown-in dislocations as shown in Fig. 16.

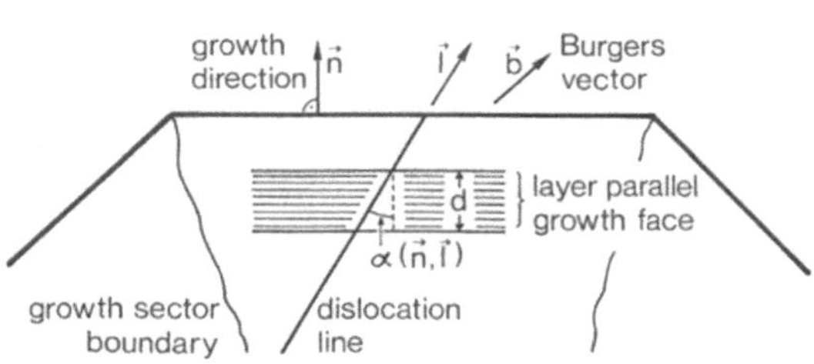

Fig. 18. Derivation of the energy of a straight dislocation within a layer parallel to the growth face

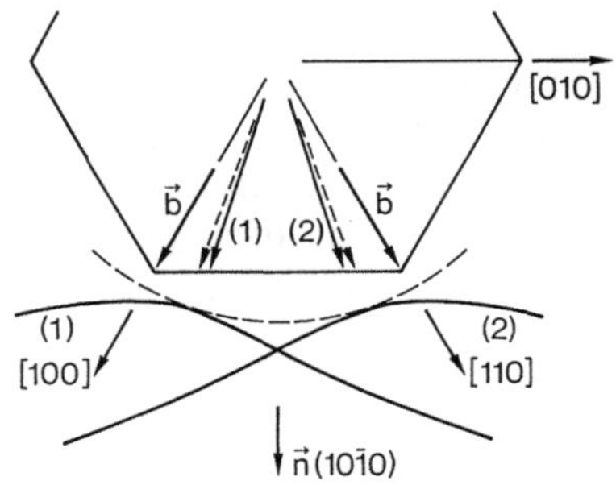

Fig. 19. Hexagonal prism of a benzil crystal and preferred directions (1) and (2) of dislocations with $b=[100]$ for (1) and $b=[110]$ for (2) in the (10$\bar{1}$0) growth sector. *Full arrows:* calculated directions. *Dashed arrows:* observed directions (deviation 1°). Curves (1) and (2) show the corresponding calculated plots of $W=E/\cos\alpha$ versus line direction in polar coordinates. The minimum values of W are given by the radius of the *dashed circle*

The preferred directions can also be explained by considering the force ("image force") which the (growth) surface exerts upon the dislocation line. This force depends on the direction of the dislocation line at the surface. It can be shown that there is always a direction for which this force is zero in any depth below the surface. (This direction is in general *not* normal to the growth face). It is reasonable to accept that during crystal growth a dislocation line follows this direction of zero force. Both, directions of zero force and of minimum energy per unit growth length are identical (see Ref. [27], p. 139).

Preferred directions have been calculated on the base of the minimum-energy approach, taking into account elastic anisotropy (for details see Ref. [33] or [27], p. 140) for dislocations of different Burgers vectors in various crystals (e.g. benzil [35], thiourea [31], ammonium hydrogen oxalate hemihydrate [32], KDP [29], pentaerythritol tetranitrate [37]). Though only the energy of the long-range dislocation strain field has been considered, neglecting the dislocation core energy, the agreement of observed and calculated directions is in most cases satisfactory or even excellent. An example is presented in Fig. 19 (benzil).

Finally, it is noted that preferred directions may be very useful for the determination of Burgers vectors, in particular in such cases where the **g·b** criterion is not successful. This has been discussed in detail by Klapper and Küppers [32] and successfully applied to dislocations in benzil and benzophenone. A comprehensive Burgers vector determination in crystals of the explosive pentaerythritol tetranitrate (PETN, $C(CH_2ONO_2)_4$) using preferred dislocation directions has been reported by Halfpenny et al. [37].

8.1.3 Post-Growth Motion and Glide Dislocations

Frequently, dislocations exhibit irregular geometries which cannot have developed during growth. An example is presented in Fig. 20, showing benzil grown from an undercooled melt. The curly dislocation lines are obviously grown-in, because they start from the seed crystal and reveal some straight segments of preferred directions. Those curly lines containing short segments parallel to the growth face must have formed after growth by glide or climb. This is explained as

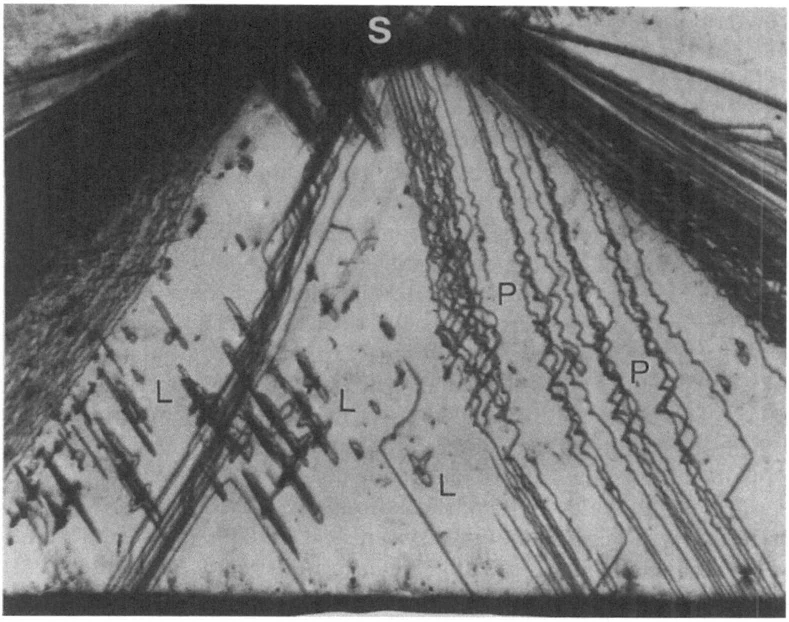

Fig. 20. Section of a (0001) plate of benzil grown from an undercooled melt, $12 \times 15\,mm^2$, thickness 1.2 mm. S: Seed crystal; P: grown-in dislocations showing post-growth motion; L: glide dislocation loops around small inclusion (bubbles)

follows by considering the curved dislocation line in Fig. 21. It is obvious that this geometry cannot develop when the growth front advances from A–A to A′–A′. A dislocation line cannot proceed against the growth directions as it does along the segment 1. The segments 2 and 3 parallel to the growth face, too, cannot have been incorporated during growth due to the forces which the surface exerts on dislocations. These "image" forces are infinitely high in the surface and drive a parallel segment out of the crystal. Thus, line segments parallel to the growth faces or "reversed" with respect to the growth direction are indicators of post-growth motion. It should be noted that a motion forming such segments can occur already during growth at distances behind the growth front where the image forces of the surface are small.

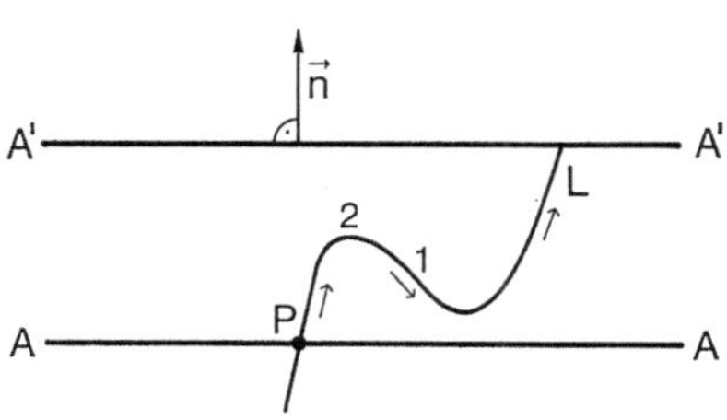

Fig. 21. Dislocation trajectory which cannot have developed during growth

Post-growth motion of grown-in dislocations occurs either by glide induced by mechanical stress or by climb due to the absorption of vacancies or self-interstitials. In benzil crystals grown from an undercooled melt, motion by glide is frequent. Motion by climb has been presumed for some dislocations in crystals of the explosive cyclotrimethylene trinitramine (RDX, $(CH_2\text{-}NNO_2)_3$) by Halfpenny et al. [38–40], who observed by X-ray topography helical dislocations following and encircling the direction of the originally straight dislocation line. It is well-known that dislocations with screw components can be transformed into helical configurations by a climb mechanism.

The above criterion for post-growth motion, i.e., the occurrence of dislocation line segments normal and "reversed" to the growth direction, implies that dislocation loops or half-loops are created always after growth. Such loops are emitted from stress centers, like inclusions. They are always glide dislocations inducing stress relaxation by plastic deformation. Examples are shown in Fig. 20. The small inclusions emitting the loops are in some cases indicated by small contrast spots. The Burgers vectors of these glide dislocations are parallel to one of the three shortest equivalent lattice translations $\langle 100 \rangle$ within the hexagonal plane (0001), which is the glide plane (glide systems $\langle 100 \rangle$ (0001)). Note that some of the loops are elongated in the direction of their Burgers vector (glide direction), thus forming long segments of pure screw dislocations.

Another example of glide dislocations in benzil will be presented in Fig. 31. Dislocation loops emitted from inclusions and other stress centers were observed in solution-grown crystals of diacetylene (PTS) [41] and tetraoxane [42].

8.2 Crystals Grown with Curved Interfaces (Czochralski Crystals)

Curved growth faces appear in almost all methods of melt growth (except for growth from undercooled melts) and in special techniques of condensation from the vapor phase. Concerning organic molecular crystals, however, only a few X-ray topographic studies on sufficiently perfect crystals revealing the typical configuration of growth defects were reported [53–55]. The crystals were grown by the Czochralski method.

In such crystals dislocations are also nucleated at inclusions which frequently are generated in the zone of first growth on the seed crystal or by serious fluctuations of growth conditions. Sometimes dislocations also originate from steps at the "crystal cone" during growth with increasing diameter. The dislocations propagate with directions (approximately) normal to the interface. This corresponds to the tendency to minimize the dislocation energy per unit growth length as described for habit crystals in Sect. 8.1.

Due to the alignment normal to the growth front, the configuration of grown-in dislocations and the resulting crystal perfection depend strongly on the shape

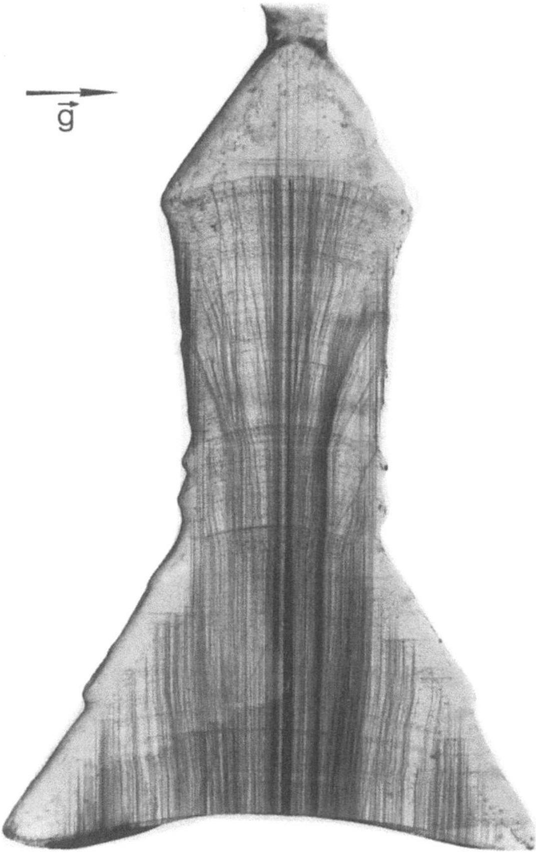

Fig. 22. (010) Plate of a Czochralski-grown salol crystal ($T_m = 42°C$), 1.3 mm thick, 43 mm long, pulling direction [100]. The topograph was taken in the reflection 002 (g is the diffraction vector)

of the interface. For convex (towards the melt) interfaces the dislocations diverge and vanish during growth through the side surface of crystal boule. This effect is widely applied in the Czochralski growth of dislocation-free crystals. If the interface is concave, the dislocations are focussed into the center of the boule, leading to a progressive degradation of the crystal quality.

An example of this dislocation behavior in organic Czochralski crystals is presented in Fig. 22 (salol, $C_{13}H_{10}O_3$). The concave growth interface is marked by numerous tiny bubbles of dissolved (atmospheric) gas which segregated during growth fluctuations. These bubbles are the source of dislocations which align normal to the interface and are focussed towards the boule axis. The increase of the dislocation density to the end of the boule is apparent. Another example is shown in Fig. 34.

9 Characterization of Growth-Sector Boundaries

In principle, growth-sector boundaries (cf. Sect. 2) are present in all crystals grown on interfaces with abrupt changes of growth directions. Frequently there are no lattice perturbations along the boundaries which then are not directly visible on X-ray topographs (e.g., in benzil crystals grown from purified undercooled melts). In such cases the boundary can be recognized by the bends of the grown-in dislocation lines (cf. Fig. 14). In organic crystals grown from solutions or from impure undercooled melts, "faulted" growth-sector boundaries, accompanied by lattice distortions, are common.

On projection topographs, faulted growth-sector boundaries appear by kinematical contrast (extinction contrast) due to lattice distortions, or by dynamical fringe contrast similar to that of stacking faults. Boundaries generating kinematical contrast are found in Fig. 27, whereas an example of a fringe-contrast boundary is presented in Fig. 23. Frequently, both kinds of contrast appear in combination, as is shown in Fig. 24. The double kinematical contrast originates from the emergence of the boundary at both plate surfaces and will be explained below.

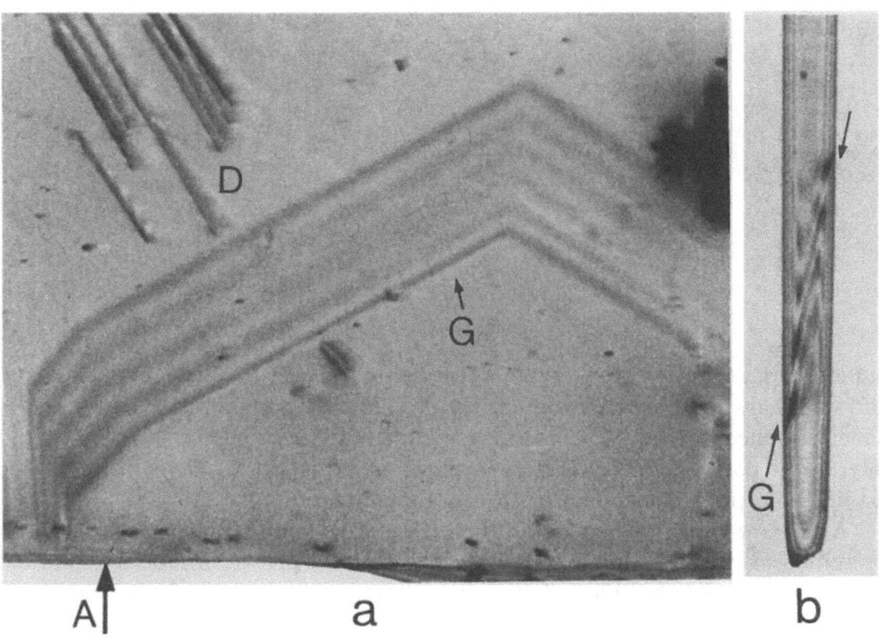

Fig. 23. Benzophenone ($1\bar{1}0$) plate, 1.2-mm thick, section $8.5 \times 11\,mm^2$, reflection 220; G: growth-sector boundary; D: dislocations penetrating the plate. At the *right* a section topograph is shown (higher magnification), taken at the location marked by the arrow A (after Götz [51])

a

b

c

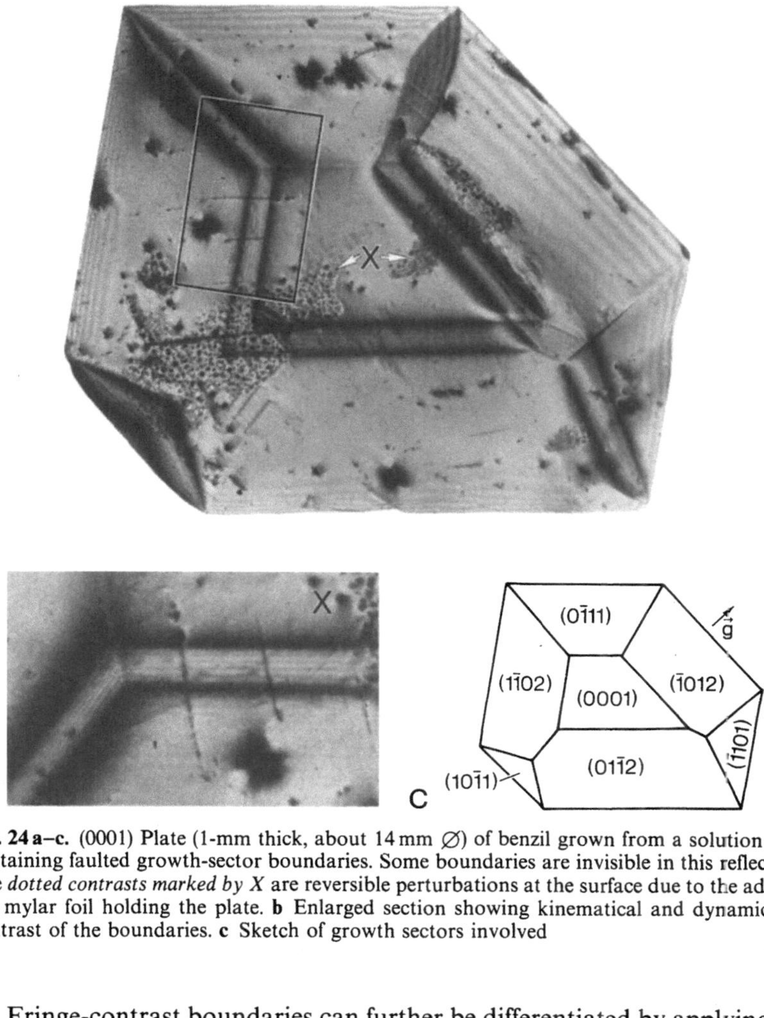

Fig. 24 a–c. (0001) Plate (1-mm thick, about 14 mm ∅) of benzil grown from a solution in xylene, containing faulted growth-sector boundaries. Some boundaries are invisible in this reflection 1012. The *dotted contrasts marked by X* are reversible perturbations at the surface due to the adherence of the mylar foil holding the plate. **b** Enlarged section showing kinematical and dynamical (fringe) contrast of the boundaries. **c** Sketch of growth sectors involved

Fringe-contrast boundaries can further be differentiated by applying section topography. The following cases can be distinguished [34, 43–45]:

a) The lattices on both sides of the boundary are exactly parallel, i.e., the perfect-crystal reflection curves coincide. In this case, due to the "interbranch scattering" of dynamical wave-fields at the boundary, a characteristic interference pattern ("hour-glass pattern") typical for stacking faults appears on section topographs [18]. An example is shown in Fig. 25a.
b) The lattices have (slightly) different interplanar spacings or are tilted with respect to each other in such a way that the perfect-crystal reflection curves do not overlap, i.e., those X-ray waves "reflected" on one side of the boundary will pass on the other side through the crystal without resonance, and vice versa. Since the effective tilts (i.e., the tilt angles within the incidence plane) are

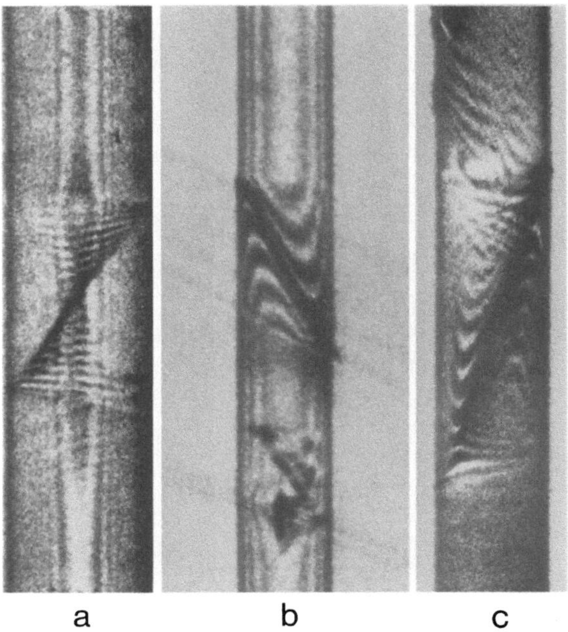

Fig. 25 a–c. Section patterns of various growth-sector boundaries in benzil. **a** Coincidence of the reflection curves of both sectors involved, (shift boundary). **b** Separation (no overlapping) of reflection curves (tilt boundary). **c** Partial overlapping of the reflection curves (tilt boundary)

usually much smaller than the primary-beam divergence, reflection is simultaneous on both sides of the boundary. The section pattern is that of two complementary wedges attached to each other (Fig. 25b). From this pattern it can be derived that the angular difference of reflection positions (which generally implies lattice tilts and changes of interplanar spacings) across the boundary is greater than about $2w$ (w: perfect-crystal reflection range, cf. Table 1, Sect. 6.3). In the case of Fig. 25b (reflection 0003) the widths of the perfect-crystal reflection range is $w = 0.67''$, i.e. the effective tilt is greater than $1.34''$. The growth-sector boundary in Fig. 23a is a tilt boundary of this type, too, as is recognized from the section topograph Fig. 23b.

c) The effective lattice tilts and the changes of interplanar spacings are such that the reflection curves on both sides of the boundary partially overlap. Now a rather complicated interference pattern results, as is shown in Fig. 25c. The effective tilt of the reflecting planes is smaller than $2w$, i.e., of the order of a few seconds of arc. In the case of Fig. 25c, the apices of the fringes converge to two points inside the section topographs close to the margins, thus forming a pattern which is more similar to case (b) than to (a). From this an only slight overlapping is concluded. The effective tilt is estimated to be roughly $1.5w = 2.4''$ ($w = 1.57''$ for reflection $2\bar{2}00$). Section interference patterns of case (c) are studied and explained in detail by Sauvage and Authier [43].

In case (a) the boundary is assumed to consist of a thin layer of impurities, introducing a shift of the exactly parallel lattices on both sides of the boundary by a fault vector. The tilts and changes of interplanar spacings in the cases (b) and (c) result from differences in the impurity content of the two growth sectors. The difference in lattice parameters may also produce a distorted transition layer which gives rise to kinematical contrast. Such contrast frequently appears at the outcrops of the boundary in the specimen surface because here, due to stress relaxation, stronger distortions are present than in the interior. This explains the double contrast of growth-sector boundaries as shown in Fig. 24.

The above results and explanations demonstrate that the planar defects "growth-sector boundaries" exhibit a variety of features, from stacking fault patterns to predominant distortion contrast. "Shift boundaries" (a) and "tilt boundaries" (or "misfit boundaries" [45]) (b) and (c), which appear on projection topographs by similar patterns of parallel fringes, can be distinguished only by section topography.

10 Preparation of Samples

Organic (molecular) crystals consist of molecules which are held together by weak Van-der-Waals and hydrogen bond forces. The lattice energy is low. Consequently, the crystals exhibit low melting temperatures, relatively high vapor pressures and low mechanical strength. Due to the low strength, the surface of the crystals is damaged already by slight contact with harder materials, for example by touching with tweezers. Even wiping with a cotton ball produces scratches which generate considerable X-ray topographic contrast. Inhomogeneous temperature changes of bulky crystals may induce cracks. Thus, the crystals must be prepared and handled with extreme care.

The following preparation procedure has proved to be adequate for organic crystals such as benzil, benzophenone, and salol: The crystals are cut into plates of various orientations and 1–3 mm thickness using a string saw and xylene as solvent. The solvent is diluted with petroleum in order to avoid too rapid dissolution. Solvents with high heat of evaporation, producing a strong cooling effect (such as acetone), are not suitable because they induce cracks. The grooved cut face is flattened by polishing on a soft tissue soaked in xylene. Final etching is performed by dipping the specimen a few seconds into xylene diluted with petroleum. During this procedure the crystal is carefully held only at its edges without any touching of the plate surface. After the final etching the solvent is splashed off with distilled water.

Immediately after this procedure the finished specimen is suspended between two elastic foils stretched over two rings and placed on the topography camera. This fixing of the crystal, however, is not without problems because dust particles can generate contrast spots on the topographs if pressed by the foils against the crystal surface (e.g. Fig. 13). The number of such defects can be reduced to an acceptable level by careful and clean working. A few surface damages, however,

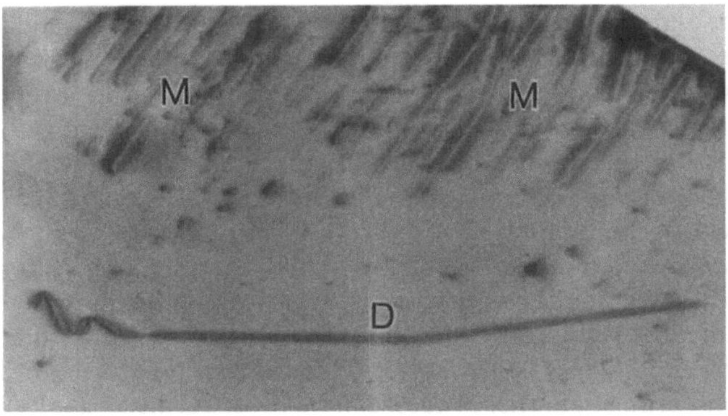

Fig. 26. Benzil ($11\bar{2}0$) plate, with a single dislocation D and mechanical surface defects M (scratches). $5 \times 9\,mm^2$

will always remain and appear on the topographs. With some experience they can easily be distinguished from bulk defects such as dislocations and inclusions (cf. Fig. 26).

The plates were preferably taken from the center of the crystals, thus containing part of the seed crystal. Topographs of such plates reveal the whole growth history, beginning at the seed in the center and ending at the edges of the plates. Czochralski-grown crystals were cut lengthwise through the center from the seed to the end of the boule.

11 Selected Examples

For organic molecular crystals only relatively few X-ray topographic studies of defects are reported in the literature. Most of them deal with solution-grown crystals, such as benzil [34, 44], thiourea [30], salol [46], hexamethylenetetramine [47], tetraoxane [42], diacetylene (PTS) [41], azobenzene [48], pentaerythritol tetranitrate [36–38], and cyclotrimethylene trinitramine [38–40]. Perfection studies of melt-grown crystals were carried out on benzil [11, 49, 50], benzophenone [50, 51], and salol [50, 52] grown from an undercooled melt, on benzil [53], benzophenone [53–55], and salol [54, 55] grown by the Czochralski technique, and on azobenzene grown by the Bridgman method [48]. An X-ray topographic investigation of vapor-grown anthracene was reported by Michell [56].

In what follows some selected X-ray topographs showing the typical defect structures of organic molecular crystals grown by various techniques are presented.

11.1 Solution-Grown Crystals

Figure 27 shows the topographs of two (0001) plates cut from the same benzil crystal which was grown by solvent evaporation at about 40°C from a solution in xylene. Plate (a) was cut through the seed crystal. Grown-in dislocations originate from liquid inclusions in the region of first growth on the seed crystal and

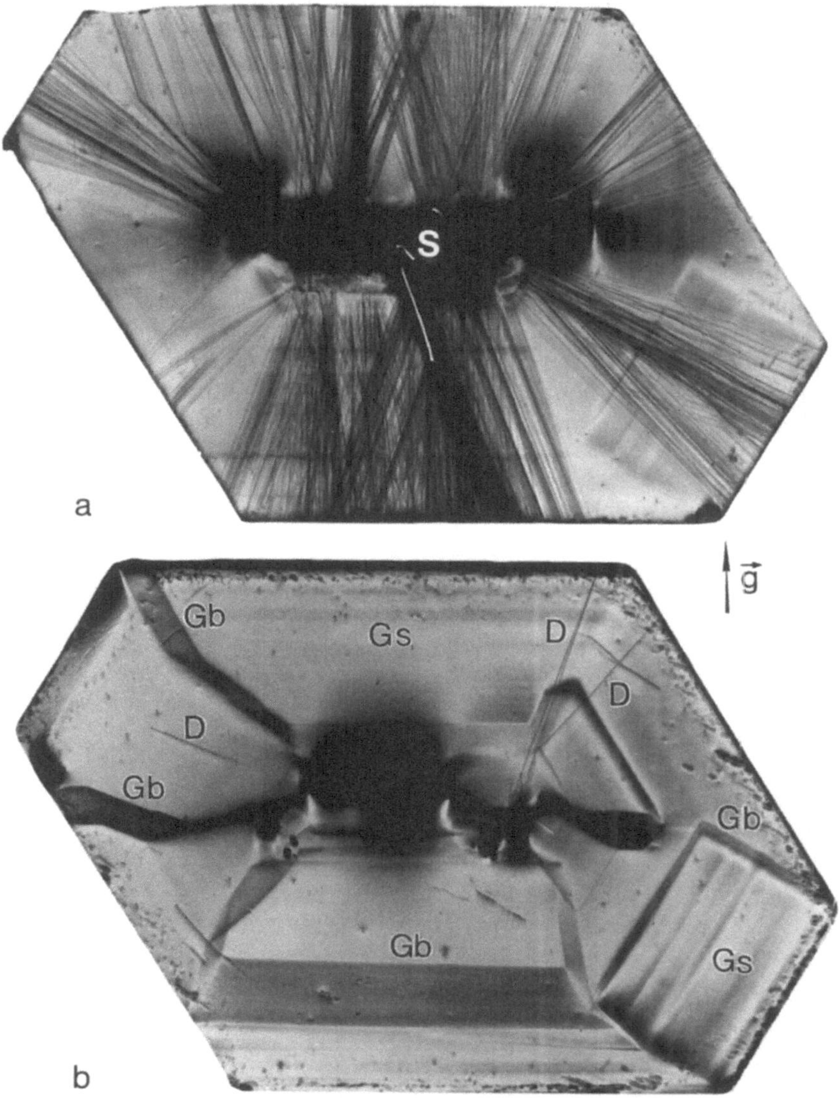

Fig. 27. Two (0001) plates of a benzil crystal grown from solution. Thickness 1.2 mm, horizontal dimension 28 mm. S: seed crystal: D: dislocations; Gs: growth striations; Gb: growth-sector boundaries

propagate with the growing faces as straight lines with the sharply defined "preferred" directions explained in Sect. 8.1.2. The Burgers vectors of the dislocations in Fig. 27a were determined by the **g·b** criterion and by analysis of the preferred directions. They are the shortest translations ⟨100⟩ in the (0001) basal plane.

Solution-grown benzil exhibits pronounced growth striations and faulted growth-sector boundaries. These defects are better recognized in the topograph of plate (b), which just "touches" the seed crystal and contains only 5 on 6 dislocations. The growth-sector boundaries, which represent the "paths" of the crystal edge between neighbouring faces during growth, appear by kinematical contrast. This indicates that they are accompanied by strain. In other cases they may generate dynamical fringe contrasts similar to those of stacking faults (cf. Sect. 7).

Growth striations consist of layers with varying impurity concentrations, which induce slight variations of lattice parameters and inhomogeneous strain. The impurities are probably solvent molecules. No growth striations are observed in benzil crystals grown from purified undercooled melts (see below).

Growth striations have also not been observed by X-ray topography in benzophenone crystals grown from a solution in xylene. Growth-sector boundaries sometimes appear by faint kinematical or dynamical contrast. In contrast to benzil, benzophenone appears to be plastic at room temperature as is apparent

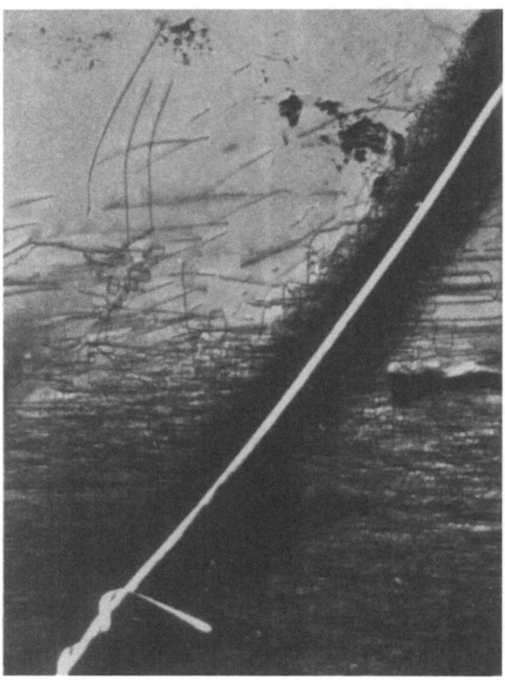

Fig. 28. Section of a (110) benzophenone plate, containing the platinum wire used for suspending the seed crystal. Reflection 002, 5.3 × 3.8 mm^2

from Fig. 28, which shows the section of a (100) plate containing the platinum wire with which the seed crystal was suspended. Due to the quite different thermal expansions, high stress developed around the wire during cooling down to room temperature. By stress relaxation many glide dislocations were generated.

11.2 Crystals from Undercooled Melts

The topograph of Fig. 29a shows a (0001) plate of benzil grown from an undercooled melt. The location of the seed crystal and the arrangement of the 6 prism growth sectors $\{1\bar{1}00\}$ is sketched in Fig. 29b, which, in addition, indicates how the crystal was suspended in the melt. The plate contains the platinum wire and loop (in the seed crystal). The absorption shadow of the wire is obscured due to the very high intensity (i.e., high local overexposure leading to "contrast diffusion") originating from the strongly disturbed neighboring regions.

The excentric location of the seed crystal and the very different sizes of the prism growth faces indicate that the growth velocities of the faces involved were quite different. The small face pointing downward was the fastest. Two reasons for this "dispersion" of growth velocities are recognized:

i) In undercooled melts the growing crystal is exactly on the crystallization temperature and thus warmer than the surrounding melt. This is due to the liberation of crystallization heat at the growth faces. As a consequence, a convectional "upward"-flow of the melt around the crystal is induced. On ascending, the melt picks up the crystallization heat. It has its highest temperature at the upper end of the crystal, spreads radially at the surface, sinks down at the cooler walls of the growth vessel, and rises up again from the bottom region of the melt in the center of the vessel, as is indicated by the arrows (Fig. 29b). Thus, the downward face of the crystal is exposed to a cooler flow of the melt than the other faces and grows faster.
ii) Numerous dislocations which are concentrated in a dense bundle and end on the small downward face promote the growth speed of the latter. Many of them passed into the neighboring prism sectors as is recognized from their abrupt bends into other directions. From these bends the development of the downward face during growth and the growth-sector boundaries involved can be reconstructed (Fig. 29b), showing that at the growth stage, indicated by the dotted line, the small face nearly vanished. At the end of the growth period the downward face grew slower and became larger (relative to the neighboring faces), obviously due to the loss of growth-promoting dislocations into other growth sectors. This example demonstrates how X-ray topography allows to reconstruct the "growth history" of a crystal.

Figure 30 shows another (0001) plate of benzil cut through the seed crystal. In this case the trigonal axis was vertical in the melt, and all prism faces were exposed to quite similar melt flow conditions. This resulted in a rather isotropic

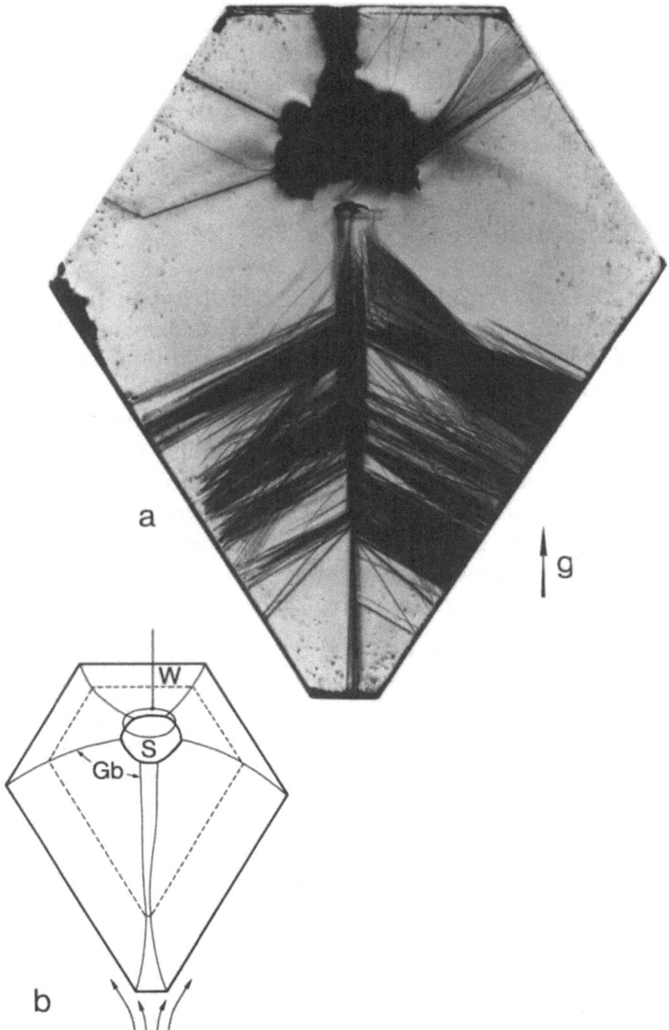

Fig. 29 a. (0001) Plate of benzil grown from an undercooled melt, 1.4-mm thick, vertical dimension 38 mm. W: platinum wire for seed crystal suspension; S: seed crystal, containing the loop of platinum wire. **b** Sketch showing the configuration of the growth sectors. The small downward prism face nearly vanished at the growth stage marked by the *dashed line*

growth within the (0001) plane. The original grown-in geometry of straight dislocations is partially destroyed by post-growth motion in the glide systems ⟨100⟩ (0001). Some dislocation lines emerge through the plate surface.

In order to demonstrate the drastic reduction of the grown-in dislocation density at larger distances from the seed, another (0001) benzil plate cut about 25 mm off the seed, containing only rhombohedral growth-sectors $\{1\bar{1}01\}$ and

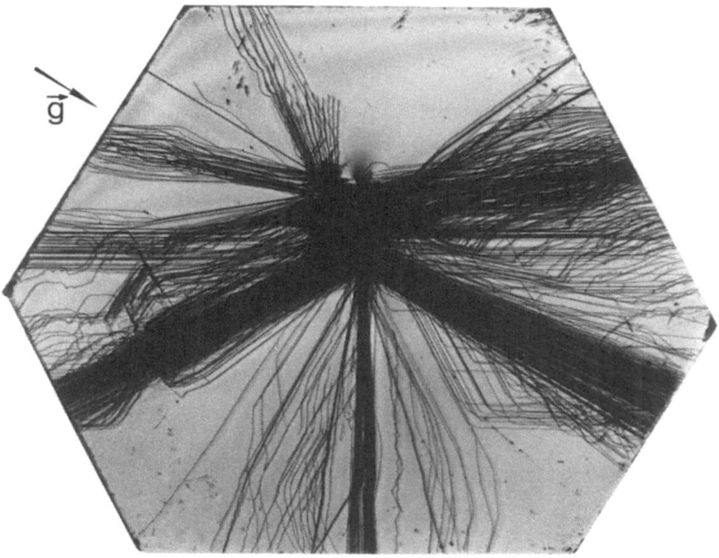

Fig. 30. (0001) Plate of benzil, 1.6-mm thick, horizontal dimension 35 mm, grown from purified melt. Many of the grown-in dislocations show post-growth motion

$\{\bar{1}102\}$, is presented in Fig. 31a. No grown-in dislocations and faulted growth-sector boundaries are found in this plate cut from a crystal which was grown from a purified undercooled melt [50]. The only perturbances result from two gas bubbles which are the source of glide dislocation loops of the glide systems $\langle 100 \rangle (0001)$ [11]. Of course, the bubbles are also the origin of grown-in dislocations which, however, are steeply inclined with respect to the (0001) plane and thus not visible in the plate of Fig. 31a.

Dislocation glide was observed only in melt-grown benzil which is plastic at elevated temperatures close to the melting point, $T_m = 96°C$. Benzil is brittle below 60°C, thus dislocation glide was not observed in solution-grown crystals.

11.3 Bridgman Crystals

Due to high stress exerted by the wall of the (glass) ampoule, Bridgman crystals are usually of bad structural perfection. X-ray topographic studies of azobenzene crystals grown by this method were performed by Kramm-Glade [48]. Despite poor quality, many glide dislocations were observed in regions of better perfection. An unusually perfect Bridgman crystal grown by Karl and co-workers is presented in Figs. 32 and 33, showing topographs of two lamellae of monoclinic 2,3-dimethylnaphthalene [57] cleaved from the same boule. The nucleation zone

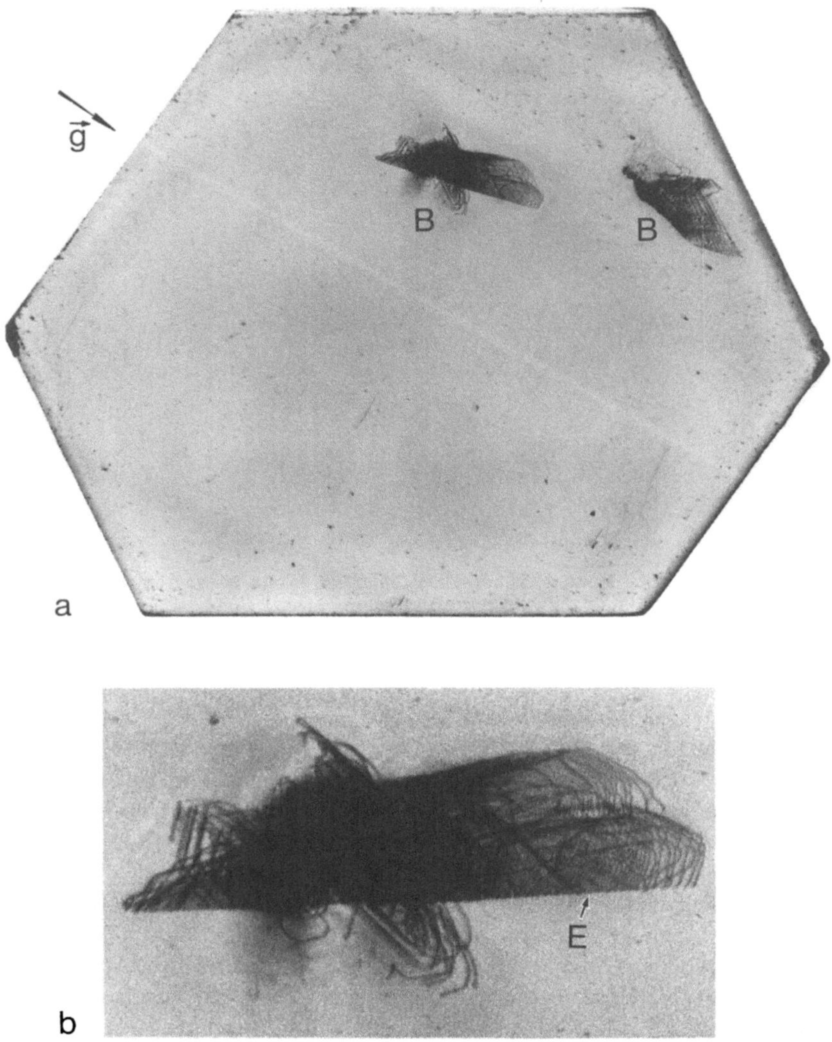

Fig. 31 a. Approximate (0001) plate of benzil grown from an undercooled melt, cut at about 25-mm distance from the seed crystal. No grown-in dislocations are visible. B: bubbles surrounded by glide dislocations. **b** Enlarged section of (**a**) showing glide dislocations ⟨100⟩(0001) emitted from a bubble. Due to a miscut of the plate by about 15° the glide dislocations emerge along E through the surface

at the left-hand side is highly perturbed, emitting half-loops of glide dislocations (stress relaxation). The right-hand part of the lamella, Fig. 32, is very perfect and contains only a few dislocations. Many more glide dislocations, emitted from various stress centers and densely accumulated in parallel half-loops (same Burgers vectors), are present in the lamella shown in Fig. 33a, b. The glide plane is always the (001) cleavage plane which has the largest interplanar spacing. It is

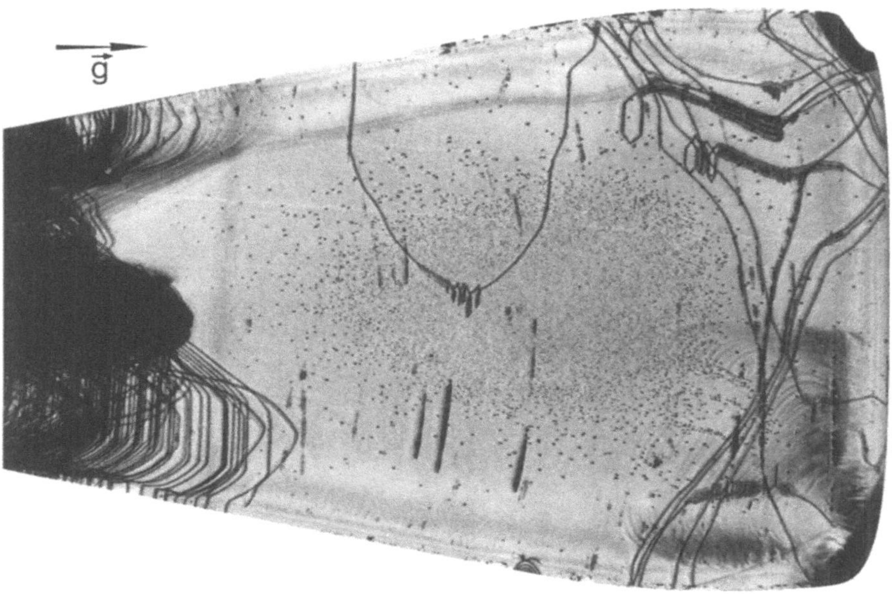

Fig. 32. (001) Cleavage lamella of 2,3-dimethylnaphthalene (monoclinic), 1 mm thick, maximum vertical dimension 17 mm; reflection 200. Using a razor blade, the cleavage was initiated from the right lower corner from which many dislocation half-loops with Burgers vector b = [010] are emitted. These dislocations are "extinct" (**g**·**b** = 0) and appear by residual contrast only

noted that the kinematical contrast of the glide dislocations, despite their mainly mixed character and the low crystal symmetry, obeys the **g**·**b** criterion (cf. Sect. 7.2) rather good, providing only faint residual contrast for reflections with **g**·**b** = 0. Thus, the Burgers vectors could easily and unambiguously be determined, e.g., as [100] (label A) and [010] (label B). More details on the defects of this interesting material will be published elsewhere.

11.4 Czochralski Crystals

X-ray topographic investigations of Czochralski-grown benzil and benzophenone has been reported by Bleay et al. [53]. An extensive study of Czochralski benzophenone and salol grown under various conditions and with different pulling directions was performed by Scheffen-Lauenroth [54, 55]. Some selected topographs of this work are presented and described here.

Figure 34a, b shows the topographs of a (010) cut through the [100] axis (pulling direction) of a Czochralski boule of orthorhombic benzophenone $(C_6H_5)_2CO$. From Fig. 34a it is apparent that many dislocations (mainly of edge character and with a Burgers vector along the shortest lattice translation [001]) originate from the interface A between the dislocation-free seed crystal S and the grown crystal. No inclusions or precipitates acting as dislocation sources are

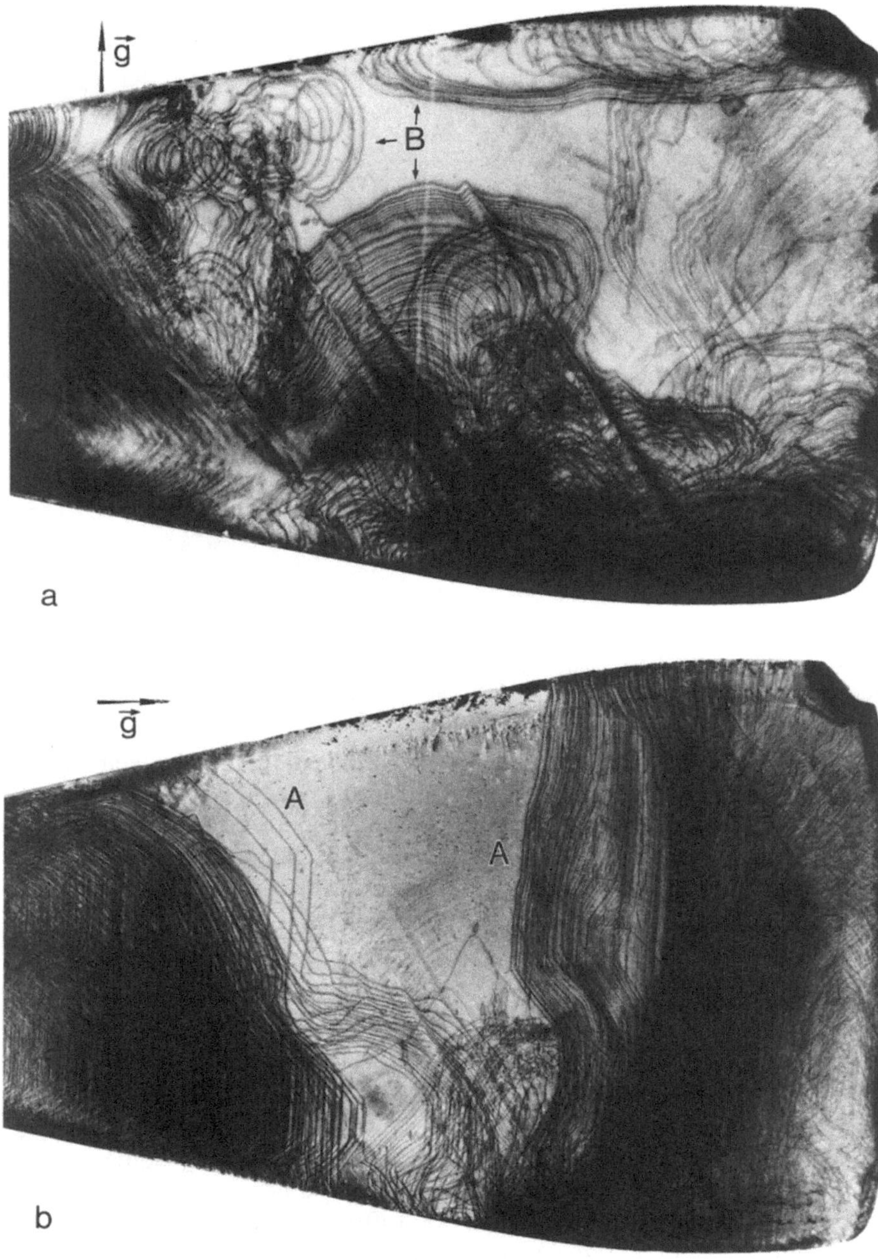

Fig. 33 a, b. (001) Cleavage lamella from the same crystal as shown in Fig. 32, 1.6 mm thick. **a** Reflection 020, **b** reflection 200. A: glide dislocations with Burgers vector b = [100]; B: glide dislocations with b = [010]

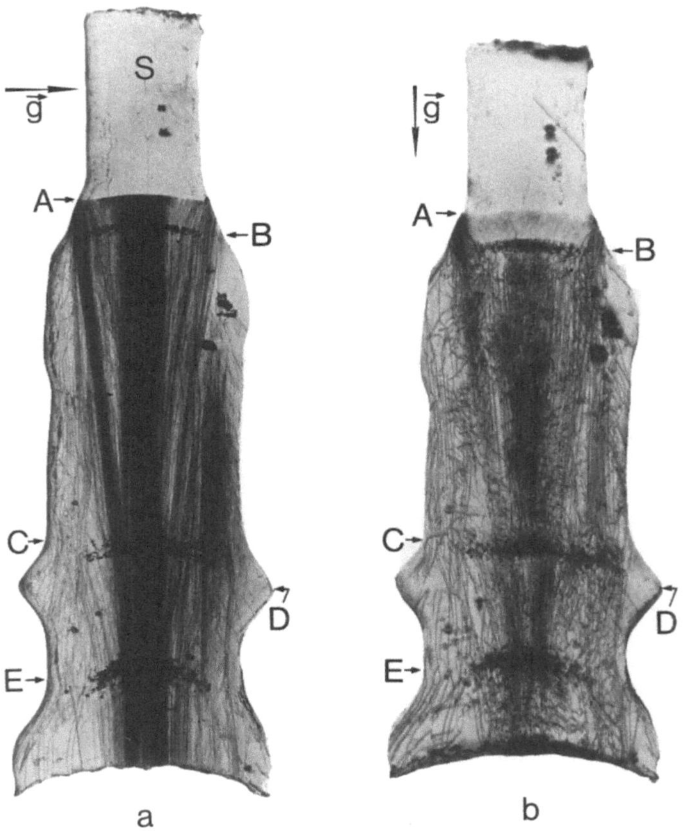

Fig. 34 a, b. Czochralski-grown benzophenone, pulling direction [100], (010) plate, 1.6 mm thick, length (incl. seed crystal S) 27 mm. **a** Reflection 002, **b** reflection 400. For the labels B, C, D, and E, see text

visually detectable along this interface, obviously because of their small size. Due to the concave (toward the melt) growth front, the dislocation lines are focussed into the boule center (cf. Sect. 8.2). The increase of the boule diameter by increasing the cooling rate of the growing crystal (cf. Scheffen-Lauenroth [54]) or lowering the melt temperature leads to a temporarily increased growth velocity and consequently to the precipitation of gas bubbles which are arranged along the actual interface and large enough to be visually recognized (labels B, C, D). No gas precipitation occurs when the diameter is reduced (temporarily diminished growth velocity, label E).

The edge dislocations appearing in Fig. 34a are invisible in Fig. 34b ($\mathbf{g} \cdot \mathbf{b} = 0$). In Fig. 34b other dislocations (Burgers vectors [100] and [010]) become visible. Due to post-growth motion, these mostly follow an irregular course. It is noted that these dislocations do *not* originate from the seed-crystal interface A (except

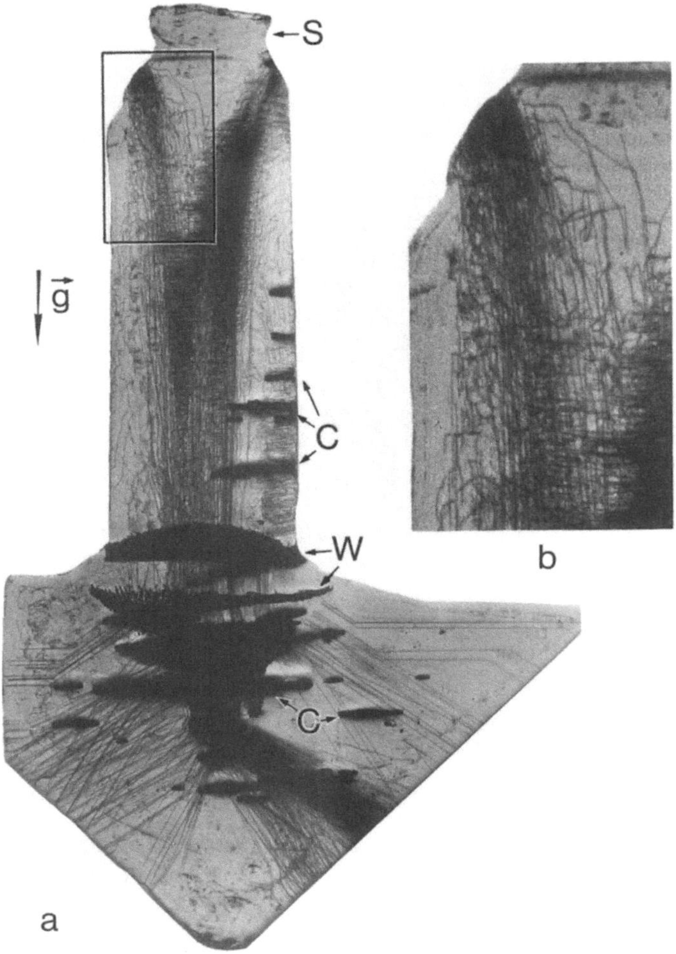

Fig. 35 a. (110) Plate, 1.5 mm thick, length 32 mm, cut from a benzophenone crystal grown partially by the Czochralski technique (pulling direction [001]), partially from a slightly undercooled melt. Reflection 002. In the transition region between the two growth modes, lenticular configurations *W* of gas bubbles have formed. S: Seed crystal, C: cracks. **b** Enlarged section from (a), showing the left cone shoulder and the irregular dislocation configuration (formed by post-growth motion)

from the edge of the seed crystal, see Fig. 34b), but from the larger gas precipitates (e.g. label **B**).

An interesting experiment is shown in Fig. 35: A benzophenone crystal was grown by the usual Czochralski technique to a length of about 18 mm (pulling direction [001]). Then a transition from growth on the concave interface to growth on planar (habit) faces was induced by stepwise lowering the temperature of the melt to slightly below the melting point (transition to growth from undercooled melt). The transition zones following the steps are filled with gas precipitates in a lens-like arrangement, see e.g. label W, with planar bottom side. The subsequent precipitation zones are partially obscured by cracks. The bubbles stop a part of the dislocations coming down from the Czochralski-grown region of the boule, as well as they act as the sources of new dislocations. These dislocations, which now emerge on planar growth faces, follow straight lines with preferred directions and are bent on penetrating a growth-sector boundary.

It is noted that in the case of Fig. 35 the dislocations (with Burgers vector [001]) visible in the Czochralski boule originate from the "shoulder" of the crystal cone (boule section with increasing diameter), but not from the interface between the seed crystal and the grown crystal. The dislocation lines are again focussed toward the boule axis. Due to post-growth motion, most of them have adopted a highly irregular geometry (see the enlarged section displayed in Fig. 35b).

12 White Synchroton Radiation Topography

The present report so far dealt with topography results obtained by applying the characteristic Kα radiation of conventional laboratory X-ray sources (sealed tubes, rotating anode generators). With the establishment of synchroton radiation laboratories a new and very powerful radiation source for X-ray topography became available. The main advantages of synchroton radiation (SR) for X-ray topography are the following:

i) A "white" spectrum of very high intensity in the interesting wavelength range between 0.5 and 2 Å is available. Very fast topography with exposure times of a few seconds or even less can be achieved. The adequate method is the Laue technique which allows the simultaneous recording of a number of reflections (topographs) in one exposure on one film. With beam diameters of 8–20 mm, sufficiently large areas of the specimen can be imaged without the time-consuming translation necessary in the Lang technique (cf. Fig. 3).
ii) Due to the high collimation of the white beam and the large distance between the source (tangent point of the storage ring) and the probe (about 40 m at HASYLAB, Hamburg, and 80 m at the Synchroton Radiation Source (SRS) Daresbury), the divergence of the beam is extremely small (<0.1 mrad). Thus a good instrumental resolution is achieved, even for large crystal-film distances which are necessary to avoid the overlapping of topographs when applying the Laue technique.

Other properties of synchroton radiation like its high degree of polarization and its time structure are of minor significance for routine SR topography.

Synchroton topographs of molecular organic crystals (benzil), recorded with a wavelength λ_0 corresponding to the characteristic CuKα line, exhibit somewhat more diffuse defect contrasts than conventional CuKα topographs [60]. This is due to the inclusion of higher-order reflections with higher-harmonic wavelengths ($\lambda_0/2$, $\lambda_0/3$, etc.). The "higher-order" topographs are (exactly) superimposed to the wanted first-order topograph (cf. Herres and Lang [61]). Since shorter wavelengths and smaller structure factor moduli (higher-order reflections have, in general, smaller F-moduli) lead to wider kinematical images (cf. Sect. 6.4), a more diffuse contrast pattern results. As has been pointed out in Sect. 6.4, long wavelengths corresponding to the characteristic CuKα line are adequate to generate narrow kinematical contrasts and thus good spatial image resolution of defects in organic molecular crystals. In this case, assuming a short-wavelength limit of the SR spectrum of 0.4 Å, higher-order reflections with wavelengths $\lambda_0/2$ and $\lambda_0/3$ are included. For an inorganic crystal imaged with an adequate wavelength of about 0.7 Å (corresponding to the characteristic MoKα line), a higher-order contribution and, therefore, a contrast diffusion does not exist. This shows that this "harmonic contamination" is more serious for the topography of organic molecular crystals.

The negative effect of the higher-order contributions, however, is partially counter-balanced by the larger Debye–Waller factors of organic molecular crystals. The greater Debye–Waller factors result from the larger thermal vibration amplitudes of the light atoms H, C, N, and O at a given temperature and lead to a strong decrease of the scattering power with increasing $\sin\theta/\lambda$. This means that the contribution of higher-order reflections is generally of low intensity, which is further reduced on the topographs by the lower photographic film sensitivity for shorter wavelengths. Thus the "harmonic contamination", though noticeably present, does not severely restrict the application of white synchroton radiation to the topography of organic crystals.

The high power of the white SR beam implies an increased danger of radiation damage. It is noteworthy that organic molecular crystals such as benzil do not suffer noticeable radiation damage, provided the very soft primary-beam components are eliminated by adequate filtering in order to avoid heating of the crystal: Conventional X-ray topographs of benzil recorded after SR exposures did not show a significant difference between irradiated and non-irradiated crystal regions [60]. In contrast to this observation, hydrate crystals such as lithium formate monohydrate ($HCOOLi \cdot H_2O$) and Rochelle salt (sodium potassium tartrate tetrahydrate, $NaKC_4H_4O_6 \cdot 4H_2O$) are severely damaged by SR exposures of only a few seconds [60]: Conventional X-ray topographs show a strong increase of the reflected intensity (kinematical contrast) in the SR-irradiated regions, obscuring completely the images of defects such as dislocations.

Finally it is noted that in certain crystals such as diacetylene, X-ray irradiation induces solid-state polymerization. This effect has been studied in-situ by X-ray topography by Dudley et al. [62].

13 Concluding Remarks

In this review the special features of X-ray topography of *organic molecular crystals* are emphasized. These features mainly result from the low absorption and scattering power of organic crystals which are composed of the 'light' atoms

H, C, N and O. Thus the 'soft' CuKα radiation is adequate for the transmission topography of these materials. The relatively low scattering power leads to large extinction distances and, as a consequence, to dynamical interference patterns with wide fringe spacings. For this reason, organic crystals of sufficiently high perfection are suitable for the study and demonstration of dynamical interference phenomena even under conditions of low instrumental resolution. On the other hand, dynamical interference phenomena are indicators of high crystal perfection.

The low scattering power of organic crystals implies a high sensitivity of the kinematical contrast to lattice strain. This means that the 'intrinsic' image widths of defects (instrumental broadenings disregarded) are comparatively large ($> 20\,\mu m$ for dislocations) and that, therefore, the resolution of defect images is poor. Under this aspect a very high crystal quality (including sufficiently low dislocation density) is required for useful X-ray topography. This condition is more stringent for organic crystals than for inorganic materials. It is demonstrated that organic crystals of such high perfection can be grown.

Among organic crystals there are also representatives which contain 'heavier' atoms, such as sulfur (e.g. thiourea $(NH_2)_2CS$) or calcium (e.g. Ca-tartrate). In such cases with higher absorption and wider angular reflection ranges the use of MoKα radiation may be indicated.

Acknowledgements: The author is greatly indebted to the Minister für Wissenschaft and Forschung des Landes Nordrhein-Westfalen for financial support, to Prof. Th. Hahn for critical reading of the manuscript and to R.A. Becker for the careful preparation of the drawings.

14 References

1. Karl N (1980) High-purity organic molecular crystals. In: Crystals: Growth, properties and applications, Vol 4. Springer-Verlag, Berlin Heidelberg New York, pp 1–100
2. Karl N (1989) Organic semiconductors: Purification and Crystal Growth. In: Mol Cryst Liq Cryst 171: 157
3. Lang AR (1959) J Appl Phys 30: 1748; Lang AR (1959) Acta Cryst 12: 249
4. Tanner BK (1976) X-ray diffraction topography. Pergamon Press, Oxford
5. Lang AR (1978) Techniques and interpretation in X-ray topography. In: Amelinckx S, Gevers R, Van Landuyt J (eds) Diffraction and imaging techniques in materials science, North-Holland, Amsterdam, p 623
6. Authier A (1978) Contrast of images in X-ray topography. In: Amelinckx S, Gevers R, Van Landuyt J (eds) Diffraction and imaging techniques in materials science, North-Holland, Amsterdam, p 715
7. Authier A (1977) X-ray and neutron topography of solution-grown crystals. In: Kaldis E, Scheel HJ (eds) 1976 Crystal growth and materials (Current Topics in Materials Science, Vol 2) North-Holland, Amsterdam, p 516
8. Izumi K (1984) J Japan Assoc Crystal Growth 11: 117 (in Japanese)
9. Read WT (1953) Dislocations in crystals. McGraw-Hill, New York
10. Klapper H (1987) X-ray topography of twinned crystals. In: P. Krishna (ed) Progress in Crystal Growth and Characterization, Vol 14: 367
11. Scheffen-Lauenroth T, Klapper H, Becker RA (1981) J Crystal Growth 55: 557

12. Authier A (1967) Advanc X-ray Anal 10: 9
13. Authier A (1977) Section topography. In: Topics Appl Phys, Vol 22: X-Ray Optics, p 145, Springer-Verlag, Berlin Heidelberg New York
14. Lang AR (1967) Advanc X-Ray Anal 10: 91
15. James RW (1963) The dynamical theory of X-ray diffraction. In: Solid-State Phys 15: p 55
16. Pinsker ZG (1978) Dynamical scattering of X-rays in crystals. Springer Series in Solid-State Sciences, Vol 3, Springer-Verlag, Berlin Heidelberg New York
17. Kato N, Lang AR (1959) Acta Cryst 12: 787
18. Kato N, Usami K, Katagawa T (1967) Advanc X-Ray Anal 10: 46
19. Epelboin Y (1987) Computer simulation of section topographs. In: P. Krishna (ed) Progress in Crystal Growth and Characterization, Vol 14: 465
20. Hattori H, Kuriyama H, Kato N (1965) J Phys Soc Japan 20: 1047
21. Hart M, Lang AR (1965) Acta Cryst 19: 73
22. Lang AR, Polcarova M (1965) Proc Roy Soc A285: 297
23. Klapper H (1976) J Appl Cryst 9: 310
24. Lang AR (1973) The properties and observation of dislocations. In: Crystal Growth, an Introduction, Hartmann P (ed) North-Holland, Amsterdam, p 444
25. Authier A, Petroff JF (1964) Comptes Rendus 259: 4238
26. Lang AR (1965) Z Naturforsch 20a: 636
27. Klapper H (1980) Defects in non-metal crystals. In: Tanner BK, Bowen DK (eds) Characterization of Crystal Growth Defects by X-ray Methods, Plenum Press, New York, p 133
28. Newkirk JB, Bonse U, Hart M (1967) Advanc X-Ray Anal 10: 1
29. Klapper H, Fishman Yu M, Lutsau VG (1974) Phys Stat Sol (a) 21: 115
30. Klapper H (1972) J Crystal Growth 15: 281
31. Klapper H (1972) Phys Stat Sol (a) 14: 443
32. Klapper H, Küppers H (1973) Acta Cryst A28: 495
33. Klapper H (1975) Habilitationsschrift, RWTH Aachen
34. Klapper H (1971) J Crystal Growth 10: 13
35. Klapper H (1972) Phys Stat Sol (a)14: 99
36. Halfpenny PJ, Roberts KJ, Sherwood JN (1984) J Crystal Growth 67: 202
37. Halfpenny PJ, Roberts KJ, Sherwood JN (1984) J Appl Cryst 17: 320
38. Halfpenny PJ, Roberts KJ, Sherwood JN (1983) J Crystal Growth 65: 524
39. Halfpenny PJ, Roberts KJ, Sherwood JN (1984) J Crystal Growth 69: 73
40. Halfpenny PJ, Roberts KJ, Sherwood JN (1986) Phil Mag A53: 531
41. Kondo M, Matsushita M, Izumi K, Ishizawa A (1984) Jap J Appl Phys 23: L752
42. Watanabe T, Izumi K (1979) J Crystal Growth 46: 747
43. Sauvage M, Authier A (1965) Bull Sec Franc Miner Crist 88: 379
44. Klapper H (1970) Dissertation, Universität Köln
45. Miltat J (1980) Misfit boundaries. In: Tanner BK, Bowen DK (eds) Characterization of Crystal Growth Defects by X-ray Methods, Plenum Press, New York, p 565
46. Lefaucheux L (1972) J Crystal Growth 16: 289
47. Duckett RA, Lang AR (1973) J Crystal Growth 18: 135
48. Kramm-Glade S (1978) Dissertation, Universität Köln
49. Scheffen-Lauenroth T (1980) Studienarbeit, RWTH Aachen
50. Göbbels M (1986) Studienarbeit, RWTH Aachen
51. Götz D (1981) Studienarbeit, RWTH Aachen
52. Szreder S (1978) Studienarbeit, RWTH Aachen
53. Bleay J, Hooper RM, Narang RS, Sherwood JN (1978) J Crystal Growth 42: 589
54. Scheffen-Lauenroth T (1983) Diplomarbeit, RWTH Aachen
55. Scheffen-Lauenroth T, Becker RA, Klapper H (1984) Z Krist 167: 159
56. Michell D, Robinson PM, Smith AP (1968) Phys Stat Sol 26: K93
57. Karl N (1984) Materials Science (Poland) 10: 365
58. Karl N, Klapper H (to be published)
59. Scheffen-Lauenroth T, Becker RA, Klapper H (to be published)
60. Klapper H (unpublished observation made at HASYLAB, Hamburg)
61. Herres N, Lang AR (1983) J Appl Cryst 16: 47
62. Dudley M, Sherwood JN, Ando DJ, Bloor D (1985) X-Ray topography of solid-state polymerization of PTS. In: Bloor D, Chance RR (eds) Polydiacetylene, Martinus Nijhoff, Dordrecht City, p 78

Author Index Volumes 1–13

The volume numbers are printed in italics

MIX
Papier aus verantwortungsvollen Quellen
Paper from responsible sources
FSC® C105338

If you have any concerns about our products, you can contact us on
ProductSafety@springernature.com

In case Publisher is established outside the EU, the EU authorized representative is:
Springer Nature Customer Service Center GmbH
Europaplatz 3, 69115 Heidelberg, Germany

Printed by Libri Plureos GmbH
in Hamburg, Germany